CORVUS

CORVUS

A Life with Birds

Esther Woolfson

GRANTA

Granta Publications, 12 Addison Avenue, London W11 4QR

First published in Great Britain by Granta Books, 2008

A CIP catalogue record for this book
is available from the British Library.

1 3 5 7 9 10 8 6 4 2

ISBN 978 1 84708 029 5 (hardback)
ISBN 978 1 84708 089 9 (export trade paperback)

Typeset by M Rules
Printed and bound in Great Britain by
MPG Books Ltd, Bodmin, Cornwall

For Rebecca, Hannah and Leah, and for David

CONTENTS

PART III BIRD SEASONS

Part I

BEGINNINGS

white feathered and perfect

1

Before Birds

On an evening in spring, many years ago, I was given a bird. The memory of the evening stays, although the details are lost: the weather, the month, whether it was late April or early May, everything else that might or might not have happened on that particular evening. I have an image, probably of a different evening, of spring rain, dusk in the street as I opened the door. Were it not impossible for it to have been any other season, even that would have been lost, but it had to be spring. What remains is that certain knowledge, and the bird.

As I write, the bird is behind me on her branch. From time to time she mutters, a sound softly bearing the imprint of the wind and the movement of trees, gentle approbation or comment, like the faintest creaking of an ancient door. Soon I'll hear the fine clicking of her toenails on the wooden floor as she walks across to stand by my desk.

When I lower my hand to her, she'll press into it an offering. It will be a damp morsel of bread, sodden, its substance now almost unidentifiable; this daily token of what I hope is love.

She's venerable now, a bird of wisdom and experience, but then, on that evening long ago, she had been only recently hatched. (The term seems odd. We're accustomed to birth but less so to hatching. How can one understand a creature not born but hatched?) Only weeks old, probably three or four, only weeks from the egg, from the unimaginable process of smashing and emerging, a bird-shaped creature four or five inches long, rounded and winged, warm, fast-breathing, pink skin glowing under a first veil of blackish, greyish fuzz. From the surface of her skin, pimples of pin-feathers were beginning awkwardly to erupt. The yellow frill round the edges of her beak gave her an air of benign and smiling calm. Clearly, an infant corvid, although the surprise was her eyes – un-corvid-like eyes, blue and bright and wonderfully, eagerly enquiring as she stared from the depth of the box in which she had been carried. (Like a human infant's, her eyes changed colour as she grew older, transforming from blue to warm brown. Now, her left eye's smoky, clouded by the cataract about which nothing can be done. The other, independent eye – for such are the eyes of birds – makes up in its unwavering acuity, for the deficit of the other.) At first we weren't entirely certain what she was but later, as she grew older, the grey of her cere, the length and disposition of her beak, allowed us to confirm her as a rook.

This bird had been found, as fledglings sometimes are in spring, fallen, pushed, abandoned, lost in the process of learning flight, who knows? I know only that she was still too unfeathered for flight, that the trees in the woods by Crathes Castle are high: larch and alder, oak and pine and beech, a canopy where another life exists above our own, another world where to fall is be cast irrevocably from the rich, raucous structures of corvid society, into a lone and lonely future.

'Found'. The word makes me retrospectively anxious, causes me to reflect on the serendipity which brought her from there to here, from the spring woods near the Guide camp where my daughter Bec's friends came upon her.

Spring is bird season. Thinking now, so many years later, of the ring on the bell, the cardboard box, is a vision of another time, opening the front door to the three eager, anxious girls standing on the doorstep, proffering box and contents, unsure if what they were offering was a gift or a burden.

A rook, *Corvus frugilegus*, named by Linnaeus in his great work *Systema Naturae*, a name meaning 'food-gathering'. (I had hoped it meant 'frugal', because I enjoyed the idea of her being part of the tradition of rural Scottish frugality, but apparently it does not.) I say 'her' but there was no method we knew of to identify her sex. We designated her female arbitrarily, discovering only last spring, when for the first time at the age of sixteen she laid two eggs, that we were correct. We named her, a choice probably now too prosaic, too frivolous for the dignity she's attained. It was derived from a piece in the edition of the *New Yorker* I was reading at the time, mention of a drag artist called, I think, Madame Chickeboumskaya. Thus, she is Chicken.

But for me, there is no name. All the ones I think of seem wrong.

They portray things I'm not: bird-keeper, bird-owner; 'bird-keeper' with its suggestion of zoos, of formality, lines of well-planted aviaries, a brisk person in overalls; 'bird-owner', which feels bossy and possibly custodial. It holds resonances of mills, cars, slaves. Ornithologist won't do because I'm not one, nor biologist, nor twitcher, nor birder. I'm none of those. 'Amateur domestic ethologist?' a friend suggested, but I'm not that either. If 'owner' has resonances, 'ethologist' has more. 'Ethologist' sounds, and is, more, far more than I am. It is weighty and grave, too much so for me.

Chance, a single moment, the confluence of fallen bird and receptive human, has changed me from observer to something else, something I can't even name: adoptive parent, housemate, beneficiary. Perhaps there is no name, no need for a name. A rook lives in my house and now, a young crow. I am *in loco parentis* for an elderly cockatiel. For years, a magpie, a starling, some small parrots and two canaries lived here too. A number of doves inhabit a small outhouse – more of a shed – a structure elevated from its lowly role of having once contained coal by being given the name of dove-house (or more correctly, as it's a Scottish dove-house, doo'cot). The inhabitants, an assorted bunch, columbinidae all, are mainly white. Two or three are brown. Some of the white ones have markings of petrol, grey, navy, apricot on their bodies and wings, a few have feathered feet. They fly freely and return reassuringly each afternoon or evening at dusk, according to season, accepting, with the cool aloofness of those who know it is their right, the food I provide. I don't know how many there are because I rarely count them. There are enough.

Birds have arrived, the chosen and the unwanted, the damaged, the accidentally displaced from nests. They have stayed, or gone, leaving, all of them, their own determined avian imprint, entirely unrelated to size or species, and with each has been established an enduring sense of connection, one that extends far, towards a world, a life, a society, of which once I knew nothing at all.

Of all of them, it has been the corvids, the rook, magpie and crow, who have altered for ever my relationship to the rest of the world, altered my view of a hierarchy of form, intellect, ability; my concept of time. The world we share is broad, the boundaries and differences between us negligible, illusory. That these relationships existed, and exist, surprises me no less than it does most people who know me, for nothing in my previous life, that now unimaginable birdless, pre-bird existence, presaged it, neither knowledge nor conscious inclination.

Had I ever thought of it before, I'd have judged myself an unlikely candidate (and, for the birds at least, an unwise choice), for I recognise that our respective worlds might have seemed, at best, separate, divergent: theirs sky-bound, high; mine more than just terrestrial but resolutely urban, differences I might sensibly have regarded as discrete, unchangeable, parallel for ever, beyond the point or possibility of meeting. I summon now, with shame and retrospective scorn, my past self and a display of abject panic, fear and horror on finding a pigeon dead in my mother's garden in London. I had to ask a neighbour to remove it. *A dead pigeon.* (The neighbour, a well-known actor, removed the bird with heroic air. We all celebrated the feat with the ceremonial opening of a bottle.)

I think still of the dogs I was brought up with, of their names, their presence at significant moments of my life, of the one-sided conversations, the nameless comforts of a silent listener, the inevitable experience of a child's loves and griefs. The last of them died shortly before I left home, and the death seemed to mark the end of childhood, the end of my life in one place, when I began a new one, of traveller, student, itinerant.

Now though, looking back, I see where it might have begun, the force that may or may not have influenced the future, or part of it at least. The Glasgow house where I was brought up had, in the way of houses of the time, no heating, only inadequate fireplaces, most of them unused, and a large stove in the kitchen. The house was large, built of stone and, winter or summer, glacially cold. The possibility of installing heating was, as I recall, briefly discussed, my father's reluctance to have the Arts and Crafts panelling warped by the drying effects of central heating eventually overriding all other considerations, and so we continued to endure the almost universal experience of Scottish life of the time, ice on the insides of bedroom windows, fierce dashes from bed to clothes, the only warm piece of the anatomy at bathtime being the portion submerged. In winter we did the best we could, huddled over any available source of warmth. I don't remember what we wore but I do remember that my father, concerned for their comfort and well-being, insisted that our dogs, three of them, wore sweaters indoors. They were striped and short-sleeved, and transformed them into indeterminate creatures, somewhere between bee and dog (although, being peculiarly well-fleeced poodles, their need might have been considered rather less than our own). While other people thought it odd, we didn't, which makes me wonder now if it might have contributed towards giving me a certain stout immunity to opinion. What it definitely did was prevent me from gaining any false ideas of the superior place human beings occupy in the world.

For years I moved around between countries and cities, a circumstance that prevented any thoughts of animal-keeping. My contact with creatures was tangential, incomplete, and when I think of it now, all I can summon is a set of impressions, random sights, remembered sounds that lead me back to a particularity of time and place: the voices of doves in the early days of a Galilean springtime, the cockroaches, patent black and brimstone brown, of alarming size that would appear with miraculous suddenness, apparently from the air of the flat where I lived in Jerusalem; of waking, bewildered, on one of my first nights back in Scotland, wondering where I was and what was the origin of the hellish sound that woke me, before I realised that it was a cat howling grievance from the darkness of an Edinburgh roof.

When I married David, who was going to be a doctor, I began still more years of itinerancy. Brought up in rural Zambia, David's interests lay, at least in his youth, more in wild than domestic animals. The pelt of a civet which he had stalked, shot and skinned accompanied us everywhere as we undertook what was necessary then in pursuit of a career in medicine: moving every couple of years in a cyclic routine of

selling, buying, packing up, taking the slow steps towards readjust-
ment, then beginning it all again. The remains of this unfortunate
member of the family viverridae, which we used as a rug, was an
unlovely object, vaguely sparse and balding in places, sprouting in
others with patches of coarse, spotted fur. The interest in hunting
appears, fortunately, to have expired with the civet.

During the years of moving, we didn't think of keeping anything
besides ourselves as we shifted us and chattels and eventually two
small children, Rebecca and Hannah, from one place to another. We'd
both been at university in Edinburgh and left it when David got a job
in a hospital in Lochaber, the first of the moves – some undertaken
with more anticipation, more joy than others – that would, over the
next ten years or so, take us from Fort William back to Edinburgh,
then to Aberdeen, to London and back to Aberdeen again. We seemed
to be forever poised to go, migratory, only partly related to place.

A further development on the road which led to birds, I trace to
the first time we lived in Aberdeen, to our home on the lower
two floors of a terrace house in an eighteenth-century street in the
centre of town. Behind the elegant row of houses, in smart streets
of boutiques and delicatessens, courtyards and outhouses opened
on to quiet lanes. In one of the outhouses in our back yard were
cages of rabbits kept by the couple who lived upstairs and ran an
informal boarding house for oilworkers, brickies, scaffolders and the
many other 'boys' who seemed to do nothing all day but sit in the
living room with the television, and smoke; a rag-tag lot, both human
and rabbit. The owners of the establishment were Georges, a Belgian,

ex-Foreign Legion, ex-mercenary, and Fee, his plump Aberdonian fiancée with the exotropic eye and voice of a triumphant seagull. I don't know why they kept rabbits but the enterprise was haphazard, the cleaning of the rabbit sheds sporadic. I still occasionally have dreams about those sheds, about damp straw and filthy cages, a focus no doubt of some uneasy guilt of my own about things not done and tasks neglected.

Without asking us, Georges one day gave the girls a pair of small white rabbits, which grew in a distressingly short time into large white rabbits. They were brought into the house frequently and were for the most part continent and uncomplaining, hopping and shuffling up and down the stairs, only occasionally attempting to eat the wallpaper or chew electric cables. We enjoyed their presence but when after a couple of years we moved to London, the rabbits stayed behind without any great regret being expressed, or even the kind of plaintive request that might have inspired guilt. I suppose that the proximity of so many other rabbits and daily reminders of the consequences of a combination of fecundity and lack of hygiene helped lessen their charm. We did take with us the tortoise one of the 'boys' had given to the children, a tortoise as any other, slow and ageless, named, for a reason I can't remember, Isambard. While I stayed behind to finish packing up the house, David and the girls travelled to London by sleeper, taking the hibernating Isambard, swaddled in his box, with them. When, after a few days in the back garden of our house in Kentish Town, he had shown no inclination to move, the proper diagnosis was made. (I assume that it must be difficult to tell much about

only occasionally attempting to eat the wallpaper

ectotherms, those creatures with the odd and fluctuating body tem-
peratures, the ones called 'cold-blooded', or ones with exoskeletons for
that matter, who in the usual run of things may do very little to allow
one, even one medically qualified, to differentiate between their being
alive or dead.)

In London, we lived not too far from Regent's Park Zoo. Years
before, after we had moved from Glasgow, I had lived with my mother
in a flat opposite the zoo, near Lord Snowdon's aviary, so near that at
night, through the sound of traffic on Prince Albert Road, the calls of
the taiga, the steppes, the savannah reached us, sounding through the
north London night, an experience described by Patrick Leigh Fermor
in the introduction to *A Time of Gifts*, when in the late 1920s, from his
mother's house at Primrose Hill Studios (where one door had been
marvellously decorated by her neighbour, Arthur Rackham), he could
hear the sound of lions from the zoo at night. We went there often and
it was on the journey between home and zoo that we used to pass
Palmers pet shop in Camden Town, a splendid institution above
whose door hung a sign that appeared to suggest that naturalists, as
well as monkeys and talking parrots, might be purchased within. One
day, we went in and bought a rat, Rupert, worthy representative in
both intelligence and beauty of the genus *Rattus*, who lived out his
allotted thousand days amicably enough with us, undertaking the
journey when at last we made the definitive move, the one we hoped
and expected would be the last, back to Aberdeen.

We left the edgy clamour of London and came here to Aberdeen, to
a house as unheated as my childhood home (a situation swiftly

altered), one much as all houses are in this part of the city: stern, strong edifices built from unforgiving granite hauled from the deep quarry at Rubislaw only half a mile away. Granite, in its singular density, creates a truly Protestant aspect, forbidding, grey, accepting of little ornamentation, a stone that in its greyness, its occasional, glittering brilliance, has created a city that seems to mirror a gunmetal sea, a high, ashen sky. The houses in this district are large, many of them very large, but all sufficiently plain as to avoid offending God with ostentation. It's a district that was built by granite merchants for themselves and the other wealthy of the city – lawyers, doctors, families grown rich from the fish trade – much as now, except that people in the business of North Sea oil have replaced the granite merchants, people who, to judge by the frank displays of glossy, elephantine vehicles (whose number plates seem often, oddly enough, to feature the letters O, I and L) squatting on their driveways, clearly feel no obligation to show such deference towards God's sensibilities.

Quarrying ended in 1971. The granite used in Scottish building is imported now from China. The huge quarry at Rubislaw is silent, abandoned, water-filled, surrounded by trees which are home to the sparrow-hawks that, from time to time, drop from the sky to feast on one of my doves. A colony of roe deer, which must have been there since before the barrier of stout wire was put up, lives undisturbed behind the fences that keep people out.

We moved into three storeys of dark, alarmingly varnished woodwork and staircases, a reminder in every way of the fact that it had been lived in for the majority of its hundred or so years of existence by one lady and her companion. The lady was scion of one of the famous Aberdeen granite merchants' families, who, on being widowed, moved from one house, a larger one nearby, to this one, both built from the stone her family had hewn and cut, and there are days still when I feel as if I might pass one or other on the stairs, the companion carrying buckets of coal up to the top floor where the children's rooms now lie empty, the intact fireplaces a reminder of the way people lived then, people praised as hardier than we. (I assume they were only so by necessity, because I don't see many Scots now choosing to reject the softening comforts of central heating.)

At the back of the house was a disordered vision of broken crazy paving and cramped leylandii straggling their way towards the light, a garden so neglected and misused that, for years, whenever I embarked upon rare but enthusiastic digging, I found the layers of old linoleum someone had dumped there years before, under a thin coating of soil in what were once flowerbeds. There was a scrubby lawn, some eccentric garden statuary, a shed and a coalhouse. I put on white painter's overalls and for many months breathed paint-stripper fumes and wrestled leylandii to their destruction.

I remember now the long train journey from London, the feeling of moving from clangour to quietude, the final miles, field and cliff and sea hanging at my right shoulder, the sudden view of river, of grey, spired city. It's north, 57 degrees latitude, within the ambit of the aurora borealis, that marvellous fusion of dust and sunglow which flings canopies of light across the northern skies, a city of gravity-defying, impossible horizons, of steely seas glimpsed above a line of roofs, masts at the end of city streets. The harbour jostles, clangs with oil vessels, huge, chain-hung leviathans, towering northern ferries looming over the street traffic. Built on viaducts, a busy city lives over a quiet one below of empty streets and tunnels, the dark, vaulted undersides of bridges; beyond, a hinterland of farmland, mountains, skies as high and as wide as to give a person the true feeling of the place they occupy in the universe. I always have the sense of not being from here, a sense I like, an edge of opposition, the outsider's view.

David's grandmother lived, at the time, in a village thirty or so miles south. A neighbour who kept doves asked her if she knew anyone who wanted some, and she asked us. I said I'd think about it, and did. Not only did I know nothing about doves, I knew nothing about birds. I could attach a name to a few, the most common, to fewer still a sound. I knew more about birds in literature than those surrounding me in every street and garden. My education had been partial, one-sided, left me with an odd, skewed, unlikely assemblage of knowledge, the perils of the academic degree. Usefully enough, I knew the Chinese characters for birds: snow goose, pheasant, kingfisher, phoenix. Real or fabulous, there they were and there they stayed, fixed on the pages of *One Thousand T'ang Poems*, in the lines of Du Fu and Bo Ju-yi, calling their snow-goose or pheasant calls in the neat concision of classical Chinese, with all the weight of their long and cumulative history, unending loneliness and sorrow, the loves and regrets of poets long ago. I was good with symbols. I could manage poignancy, sorrow, longing, separation. On the other hand, did I know how to feed a bird? Hold one? Had I ever looked a bird in the eye? What did I know? Nothing. Less than nothing. *yī wú suǒ zhī*. But I took them all the same.

2

The Doves

The doves arrived. I didn't then realise the momentousness of the day. The implications escaped me, the intimation that this was a beginning. They were brought from St Cyrus by Gran's unsmiling neighbour. I don't know if she was naturally usmiling or if, as I suspect, it was because on meeting me she understood that she was giving her birds into the care of an unfit person. She may have had hopes for my improvement, or even my ability to learn, as she told me in the most basic terms what to do. She demonstrated, I remember, exactly how much each dove needed to eat each day, a quantity that might have filled an egg-cup which I thought a miserly amount, an instruction to which I have never paid the slightest attention. She carried in the birds, four of them, in boxes, then left. She was, of course, entirely correct. I was an unfit person.

We had prepared the coalshed, newly designated as the doo'cot, nailing up perches and shelves, removing some bricks from the outside wall to form an entrance. Over the entrance we fixed a triangular wooden stucture with a small, arched doorway, a removable wooden hatch-door and a landing platform. We painted it white to look like a traditional dove-house. We hoped that it might provide everything a dove could want.

As soon as we had carried the boxes in and opened the flaps, the doves hopped on to the sides of the boxes and then stepped, looking with appraising interest around them, into their new establishment. They were white-feathered and perfect, and showed more confidence than I did, inspecting the interior of their house with the air of someone being shown round a dubiously appointed show-home.

Given instructions to keep the birds confined for three weeks, I obeyed. I was glad of the restriction, although it felt slightly cruel, because I was scared that when they were released they'd disappear instantly to fly back to their previous owner in St Cyrus with a volley of aggrieved complaint against me.

I thought about them constantly, worried about their welfare, whether they were eating, the state of their mental health. I looked in often to watch them, with no clear idea why, or what I would do if I discovered something untoward, or how I might tell if I did. They seemed content. They were eating. They perched. They made the sort of sounds one might have expected, their voices indicating to me for the first time that perhaps keeping doves would be more pleasure than anxiety. After a further few days, I began to creep into their house through the door at the end of the kitchen with a sense of determination, the grim, reluctant sort, knowing that I had to learn how to catch one. I'd select my victim, pursue it relentlessly, grab clumsily and, in the event of success, hold on, for a purpose of which neither of us, I think, was entirely certain.

The girls mastered the skills of dove-handling with far greater speed and confidence than I did. While I was still blundering ineffectually about the doo'cot, they'd go in and emerge calmly, holding an apparently contented dove between their hands.

On the day of eventual release, I removed the door and, after considerable hesitation and anxious hovering (mine, not theirs), they walked out on to the small platform in front of the house, looked around them, seeming almost blinded by the possibilities in front of them, and then flew. The first revelation of seeing a creature that one knows personally flying is like no other, the nearest thing one might have to flight (in the same way that one's ambitions may be fulfilled vicariously by encouraging one's children to do something they would have done anyway).

I did, in that moment, assume that this would be my final glimpse of them, but I was wrong. After some flight, a bit of brisk circling in the air above us, a period of what I assume was orientation, they flew back into their house, confirming remarkably, amazingly, the truth of the designation 'homing pigeon'. (The terms 'dove' and 'pigeon' are loose, interchangeable ones, I discovered. Doves are generally thought of as smaller than pigeons. White ones always seem, I don't know why, to be called doves.)

All birds that travel long distances – migrating birds, homing pigeons – have many methods of navigation, most of them still not entirely understood. Their knowledge of direction appears to be innate. During their first migration, birds already understand where they are to go and for what distances they must travel. Although many birds share remarkable direction-finding abilities, some species of pigeon are the most reliable at finding their way back to a precise location over distances of hundreds of miles. The way they navigate has been referred to as the 'map and compass' method, a kind of super-efficient internal GPS system that pin-points their location allied to a series of equally efficient internal 'compasses' that indicate direction; the sun and its movement across the sky is one; the stars, another. In *Gatherings of Angels*, Kenneth P. Able suggests that during the first summer of their lives birds learn the layout of the stars, including the importance of the North Star for navigation. The 'magnetic compass', the magnetoreceptor, composed of magnetite crystals situated in a pigeon's upper beak, is a mechanism that appears to allow it to find its way by sensing the earth's magnetic fields, although the precise method by which this is accomplished is still unclear. Chance, circumstance, the vagaries of weather or the activity of the sun may cause birds to become lost. Gales blow them from their course; surges of solar wind cause geomagnetic storms, distorting the earth's magnetic field, interfering with their direction-finding. (Attentive pigeon fanciers monitor daily geomagnetic activity, keeping their birds confined when conditions might be dangerous.

Strong evidence demonstrates that scent also plays an important role in direction-finding, as do visual and topographical clues, watching patterns of polarised light in the sky, following the line of rivers and motorways. Birds may be aware too of 'infrasound', those ultra-low-frequency sounds too low for the human ear, the sounds of the movements of the earth, the deep whisperings, the groanings, creakings, crackings of the fabric of universe, the sounds of sea and wind, of oceans and volcanoes, the explosion of meteors, the gathering of hurricanes far away.

As I watched my doves' light, easy flight, I was delighted by them, by their certitude and reliability, the way they flew off into the sky and returned unfailingly each evening with the onset of dusk.

The doves had lived here for a few months when the first dove-house crisis occurred. One evening, one of them failed to return. It was, I think, more likely to have been as a result of an unfortunate encounter with a sparrow-hawk than of any mishap in direction-finding, but I remember the anguish of the missing bird's distraught mate as I joined her in vigil. She watched from the roof vainly for his return while I peered from the window, ineffectually carrying out sufficient hand-wringing for us both. I didn't know what to do when the others began to peck and harass her, taking advantage, it seemed, of her solitary status. The pet shop suggested I try Kevin the bird man, owner of a small bird sanctuary north of Aberdeen. I phoned him, an ebullient-sounding Londoner who said that yes, he had several possible mates for her but he would have to see what kind of dove she was before choosing a suitable partner.

'Put her on the bus and send her up,' he said. *Put her on the bus.* We drove the eighty miles, there and back, on Cupid's errand, through the glorious green Aberdeenshire countryside. It was a pleasant journey, companionable, my dove in her box, an anticipatory empty box beside her, both of us listening to music, exchanging whatever sounds each of us thought appropriate.

I left the bird in her box in the car as we wandered around his steading, admiring the birds he already had, chatting, as one does, of bird matters. Then Kevin inspected my dove and, matchmaker *extraordinaire*, he selected for her a handsome male, white and blue and petrol, a sturdy fellow with feathered legs, the fine fringe extending to cover his toes. He was beautiful, a suitable, appropriate match, and so, booted and cooing, he was put into the back seat in the adjacent box, calling loudly.

'There's nothing,' Kevin said pensively as we closed the box, 'as randy as a male dove.'

From the first, my dove was delighted. How does one detect delight in a dove? By voice, the quality of cooing, an indefinable note of enthusiasm, an undertow of frank excitement. Listening as I drove back that day was my initiation into an undreamt-of realm of birdsong, sound, intonation. Until then I hadn't known that dove vocabulary extends beyond 'coo', that they have a wide range of tone and expression, that their voices will indicate their state of mind, from outrage to mild annoyance, from determination to protest. I didn't know that they communicate with one other in warning or in reaction, and that they appear to chat, as we might, a sound muted and domestic, like human voices heard from an open window. I began to listen for murmuring susurrations of endearment issuing from the doo'cot as pairs would sit companionably on their perches, preening each other's feathers, or for the satisfied liquid, rolling sounds at roosting time.

Descriptives such as 'coo', I began to realise, deal with one sound only. I paid attention to see if they really did make a sound that resembles 'coo'. I listened for it but 'oo' and 'woo' seemed predominant. When heard filtered down a chimney, the sound was more 'broo hoo'. In most other languages, I discovered, dove calls are rendered as 'gu' or 'ku'. There are exceptions. French doves say '*roucoulé*'. Listening to the discourse of two doves on the windowsill, I'd hear the word 'roucoule' lavishly interspersed with a variety of tones and expressions of 'ooooh'. Danish doves say '*kurre kurré*', Dutch ones '*roekoé*'. Hungarian doves say '*a galamb búg*'. (They would.) The more I listened, the more I'd hear them say everything attributed to them, whether Indo-European, Finno-Ugric, Sinitic, Semitic, Hamitic; they say it all, calling in many languages, down the chimney, from outside the study window, in the quiet of their house in evening, expressing their dislike and fears, of both cats and sparrow-hawks, their dissatisfaction with a new type of grain that rashly, once or twice, I decided to buy for them.

All the way home that day, as I drove my match-made pair, the sounds of anticipatory delight rose from both boxes, crossed the cardboard divide, a loud, mutual, trysting, lascivious cooing. Carefully I

carried both boxes in and released the birds into the doo'cot to whatever pleasures awaited them. I had to keep them all in for weeks, but
from the first their lives together were harmonious, their union obviously abundantly blessed because until now the line continues, the fine
booted legs and feet of the handsome male, the same petrol patches,
the sturdy build, the same roving, randy eye.

As they began to breed, I'd take an interest in the young, listening
for their first squeakings and cheepings, watching for their first flight,
their first return to their house, trying foolishly to help if they seemed
not to know where to go, hovering in the cold and rain for hours,
urging the small, probably quite capable bird towards the door of its
house and safety. (The rest of the family realised much earlier than I
that doves are well able to attend to their own affairs.)

I hadn't thought about anything as practical as numbers, about the
fact that, left to themselves, the doves would breed, multiply, and that
they'd do it fast. Doves are fecund. By the time I learned the trick of
egg-replacing, of removing the warm, fertile eggs and putting cold
ones in their place, there were a dozen or more.

Another thing I hadn't thought about was that they might be
capable of violence. The realisation came to me as a piercing disappointment (a bit like the day when the very small, deeply engaging
wild rabbit we'd found by the roadside and brought home looked at
me reflectively as I stroked her, and bit me). I had had no idea that
doves are, by most standards, aggressive, indulging in crimes that,
were they carried out by humankind, would have them swiftly, irrevocably, consigned to Broadmoor or The Hague. (Dove crime requires

no great forensic skill to solve, involving as it does a combination of the white-feathered and blood.)

Anything I'd read about aggression in dove behaviour seemed academic, or else I'd simply forgotten, until I found a small chick, still in its nest, being viciously pecked by an adult from another family, its head reduced to bloodied bone. The experience felt like the flight from Eden, a lesson, if any were ever necessary, in the folly of anthropomorphism, or of relying on the veracity of depiction or scope of the knowledge of natural history, of assorted Levantine scribblers and Renaissance painters.

From that day, my view of the religious art of Europe changed. There, in most portrayals of the Annunciation at least, is a depiction of the Holy Spirit in avian form, in every possible pose, guise and attitude, descending, ascending, flying, hovering, doves large and small, doves who don't look very much like doves, doves who appear to stand upon the halo or the head of Mary, doves who appear to be attempting to dive at speed towards the unfortunate woman's head (miscreant behaviour, surely, for the embodiment of the Holy Spirit). There are a few Annunciations in which the key player is notably absent. In a painting by the fifteenth-century Dutch painter Jan de Beer there is no dove, perhaps because even the Holy Spirit might be deterred by the presence of a cat sitting on the floor. (Cats, being amoral pagans, might not have the necessary inhibition to lay off the Lord's messenger.) The dove that appears in Van Eyck's Ghent altarpiece is a particularly pretty one, with rays of light expanding from and surrounding his neat white head. There they are, in so many Annunciations,

like the antlers of rutting stags

foreshortened, in rays, in beams of light, emerging from marvellous faded frescoes, soaring over cloud-borne banks of naked, levitating infants – but as far as I'm concerned they're all there under false pretences.

I learnt not only of murder, but of domestic violence, witnessing squabbles of a wilfully aggressive nature. One day I came into the house to a sound I had never heard before, a scratching accompanied by a dull thudding. Tentatively, I crept in, peered round the study door. Outside, on the ledge of the small, high, recessed window, two doves were beating the hell out of one another. They appeared to be attempting to blind one another while simultaneously endeavouring to twist off the other's head. Their perfect white wings were locked like the antlers of rutting stags as they wrestled in the small, three-sided space. Swaying, thrashing, biting, their interlaced wings struck again and again against the window. In a moment of temporary triumph, one nudged the other towards the edge of the sill until, still entwined, both plunged into the air beyond. Dipping into flight, they circled for a moment of mid-air recovery before they were back, ready for another round. Scratch, shuffle, thud. *The eyes, the eyes! Go for the eyes! The wing! Give it more wing! Whack! Twist the head! Twist the head!*

This particular dispute was territorial: both wanted to occupy the ledge above the study window. Wrestling continued for three or four minutes until the moment when, for no reason I could see, one of them flew off, defeated. The victor settled itself on the sill, until the next time. Having made its territorial gain, it tried to peck its way into the study, tapping its beak incessantly against the window.

The sound (which I'm used to now) was as ghosts might sound transporting their earthly chattels through empty rooms, creatures entombed within medieval walls, trying to get out; the inexplicably eerie beat of feather over glass, the frantic scratch and click of toenails on wood, and above it the voices of the aggrieved parties, muted by the glass, muttering bitter outrage and complaint.

They'd fight too on the low roof of the doo'cot, compelling me to rush out with a broom to try to separate the combatants, who would fly immediately away to stand patiently on the edge of the gutter, waiting until I'd gone before flying back to begin again. I used to look at the card pinned beside my desk with its depiction of Picasso's 'La Colombe sur fond noir', and think of his manifesto of peace: 'I stand for life against death; I stand for peace against war', and, with resignation, began slowly to accept that the symbol of the dove of peace convinces most when one has had minimal contact with the real thing. Sometimes I would think about the doves' egregious behaviour but the visions would collide, elide as I'd see one flying in sunlight, its wings wide, its feet curled neatly under it, its face one of sweet and beatific calm. I would forget. I accepted then that I am shallow, seduced by beauty. I knew that, when it comes to doves, this, regrettably, is the way they are.

In time, my awareness of light changed, my reaction to sound, to sudden noise and panic outside, a change in shadow or brightness, the falling-silent of the small birds at the bird feeder, the numb, heavy thud against the window. Until then, I had never been aware of the sparrow-hawks who patrol the skies above us but I learnt that even the sight of one, too high to be easily identifiable for me, would cause terror among the birds. Alerted one day by the sight of the doves flying wildly towards their house, and the sound of one colliding in panic with the window, I looked out and saw a dart-shaped brown bird rush and wrestle something on to the stones of the path. I ran out immediately to engage in brave combat with whatever it might be, armed only with a remnant from a long-ago Hallowe'en, a red-plastic trident with a wobbly pole. Devil-like, I stormed down the path brandishing the weapon towards the crouching bird who, by spreading his beautiful wings of grey-blue and soft brown, was attempting to conceal the bright white of his victim's feathers, as if by so doing, I might not notice. The dove, I could see, was alive and appeared undamaged. I jabbed the trident towards the sparrow-hawk, who turned to stare at me with a look of great, entirely deserved hatred before he rose slowly from his lost dinner and equally slowly flew away. In flight, he was exquisite. He was too, very frightening. Were I a bird, I would share their terror at the quick silhouette passing overhead like a faraway cloud. The rescued dove, unharmed, flew instantly back to its house where it sat with its friends and family, complaining for the rest of the day in a loud and outraged voice, no doubt about the need for raptor control or the unfortunate tendency of life to be unpredictable.

Always, after a dove has flown against the window in mortal fear, there remains a pattern on the glass, a faint and powdered image, a ghost imprint of outstretched wings, a small body, two smudged arcs of feathers.

I was even more anxious after that, watching for the sparrow-hawk's inevitable return. It was the final, emphatic affirmation that dove-keeping wasn't likely to make me more serene or bestow upon me a previously undiscovered ability to regard the world with equanimity. It was, I know now, the level of self-delusion that leads people to buy houses in countries they have seen only in summer, or to move from city to country believing that the essentials of life will be different there. Any sense of calm derived from the presence of the doves was transient, a series of brief interludes interspersing days of fear and panic.

The sparrow-hawk, either the one who had visited before or another, did visit again. I looked out of the window one morning. Snow in April? The flowerbed was hidden under a drift of white. In the middle, with an air of calm intent, was the hawk, plucking and tearing. The process was prolonged, the remains slight, only a pair of feet, a light scattering of feathers.

When, after we'd been here for a few years, new neighbours moved into the house next door, the first alarming sign I noticed in their garden was a large, well-kept black and white cat. The second, shortly after, was finding a dove, dead, eviscerated in a mess of blood on the dove-house floor.

The cat was well-fed, pampered and almost preternaturally feral in its instincts. For all the time it lived next door, the cat and I waged a one-sided war. All the advantages were on its side, for it was a cat, possessor of every useful, artful feline attribute, every precision implement in the armamentarium of silence and stealth. My weaponry was a water-pistol. The cat evaded and avoided every attempt at deterrence, every blast of water, my putting up elaborate barricades, installing a sonic anti-cat device, to protect my doves from harm. When I asked the neighbours if they could restrain its activities, they said in a mildly irritated way that it was only doing what cats do. For the maintenance of neighbourly harmony, I refrained from saying that I didn't care what they do; I just didn't want them doing it to my doves.

That late summer after the first murder, the doves stayed outside for days, too scared to return to their house, perching on nearby rooftops, their voices loud in warning, descending every now and again, cautiously, to feed. The weather was warm and I was happy enough to allow them to stay outside. Only when the nearby roofs began to display the consequences of their occupancy did I begin to worry about complaints. When the weather became suddenly autumnal, I stood for hours by their house, shutting them in one by one as they went in for breakfast, until they were all inside. I kept them in for a few days until the rain, for which I had waited with the anxiety of an Ethiopian farmer, washed away the consequences of summer.

The murders continued, reducing the number of doves to single figures again, and stopped only when the neighbours, and cat, moved away. The next ones brought with them several noisy, energetic children, as useful an anti-cat measure as any.

During the time of the cat depredations, of finding and removing bodies, the period of my long and futile war, I wondered if it was worth continuing to keep the doves; but I thought about what we'd all miss if we didn't have them: our pleasure in the way they looked, their presence in the garden as they lined up to bathe, wandered across the grass on damp mornings, pottered by the pond, the fanatic, obvious delight they took in flight, their luminous, stellar beauty. Their sounds had become part of our lives, their voices echoing down the sound-chambers of the chimneys, the way the movement of their wings outside altered the colours in the rooms. I considered my relationship with them too. It was, I decided, a rather one-sided matter but while the consternation, worry and appreciation were all mine, I didn't expect or hope for it to be otherwise. The doves accepted my presence. They required nothing from me but food. What I required of them was what they did, lighting winter's darkest days as flashes of white and silver against a slate or dun-coloured sky, on summer evenings, creeping up from gutter to slate to the apexes of the dormer windows on the roof as the sun lowered in the sky before heading for home, their lovely wings slap-slapping against the wind and the sky.

3

Le Jour de Gloire

Not long after we got the first doves, Rupert, the rat we had brought from London, died. We found somewhere to obtain two more. They were females, Japanese hooded rats of lissom white and charcoal beauty. We bought them not knowing that one was already pregnant, a circumstance that ensured that, for the next thousand days or so, the girls were rarely without a rat somewhere about their person, sitting on a shoulder, moving bumpily up the inside of a sleeve. We installed the offspring in strictly segregated quarters in what in other houses would be called the utility room but in ours is still called 'the rat room' in tribute to those rats, remembered still for their beauty, intelligence and charm, for their classical pantheon of names, Mars, Apollo, Venus, Aphrodite.

For her twelfth birthday not long after, Bec asked for a bird of her own, one that would live in the house. The request seemed reasonable.

What kind of bird? None of us knew. I went round pet shops, peered into cages, consulted all available works on pet birds. Finches were too small, too specialised, too unresponsive for the required purpose. Macaws and African greys were too large, too daunting, too expensive. I examined the credentials of the small grey bird with apricot cheeks I had seen in pet shops, a cockatiel, *Nymphicus hollandicus*, in spite of his name a bird of Australian origin, member of the cacatuidae, the cockatoo branch of the parrot family. The day before the birthday, the bird was bought. He was eleven weeks old. By comparison with the doves, he was tiny. If I'd been frightened of the doves, I was even more frightened of him, of his apparent fragility, of what I didn't know about how to look after him. I became suddenly aware of the dangers of open windows, the possible consequences of shutting in, of shutting out, of standing on, sitting on, of wrapping inadvertently among the laundry; a neurotic's invariable catalogue of the fears and anxieties of keeping anything, children too for that matter, although children, after a certain age, will speak, express their requirements, look for food themselves when necessary, complain.

Handing the bird to Bec on birthday morning was a relief. I already suspected that hers would be more capable hands than mine, and so they were. She took him in his cage upstairs to her room and, in a process both secret and effective, emerged with a confident, speaking, one-woman bird, a bird whose devotion to her has not flagged, a process that had the additional, unforeseen effect of engendering in him a dislike of me, a position that, with heroic adherence to his own loyalties, he maintains to this day, hissing, swearing, trying to bite me as I carry out the daily feeding, cleaning, chatting and bathing which has been my obligation, and indeed pleasure, in the years since Bec left home. (Proof of bird memory, and perhaps even intuition, if anything is, he responds with a glorious and unalloyed joy at her return, appearing, in a way I cannot explain, to know when she is en route, reverting to the song patterns, speech and cockatiel-mantra-shrieking that characterised the years when they both lived here together.)

The cockatiel, named Bardie, was a good choice, a fine, if salutary introduction to the intelligence of parrots. I did not know that any bird could imitate, with an astonishing degree of perfection, the sound of telephones, that they could both speak and use words purpose-fully. I did not know the ear-shattering power of their voices. Nor could I have imagined, before becoming acquainted with Bardie, that the anger and passion displayed by doves was not only common behaviour in birds but could be surpassed and extended by a capacity for irritation allied to frank displays of uncontrolled rage.

The process that had begun seems almost enshrined in a natural law that lays down that one bird, entirely aside from any biological con-sideration, swiftly begets more. As people began to know that we kept birds, the pace gathered momentum. Friends, schoolfriends, the par-ents of friends, neighbours, someone a neighbour had met in a shop, brought them to us, believing, it seemed, that we knew about birds because having even one bird appears to render one shamanic, appar-ently possessed of mystical knowledge, which attracts every waif and orphan, every ill, injured or dying bird, every unwanted unfortunate, every happened-upon, abandoned infant, every possibly runty faller-out-of-trees.

They arrived, and since there was nowhere else for them to go, they stayed. Most were very small: unfeathered blackbirds, tiny thrushes that, in spite of our efforts, died. They'd open their small beaks wide, form them into deep, pink pouches to receive a minute quantity of food from an eye dropper or the end of a salt spoon. I'd wipe the residue of mush from their faces and then, one day, their eyes would turn dull and they would die. I'd try to believe that it was not my fault.

I began to buy books on raising infant birds, on feeding them. I bought boxes of insect food, everything to prepare myself – for what or whom, I didn't know. So that I might do better the next time. For the next time the doorbell would ring. One of the books I bought, picked up from a bookshop sale-box, serendipitous among the out-of-date computer books, hotel guides and sundry oddities people had ordered but forgotten to collect, is called *Feeding Cage Birds: A Manual of Diets for Aviculture* by Kenton C. and Alice Marie Lint. (Kenton Lint, curator emeritus at San Diego Zoo, avatar of bird nutrition, oh fortunate man!) My copy is falling apart now, the pages threateningly adrift, so irresistible are its contents, relentless practicality matched with a soothing litany of arcana. The contents page itself is a delight:

Preface
Note on Availability of Foods

Sphenisciformes
  Penguins
Struthioniformes
  Ostrich
Casuariiformes
  Cassowaries, Emu . . .

Thus it continues, through heron and ibis, vulture and frogmouth, every bird that might or might not plummet from its nest, be blown in by a freak wind, happen by circumstance unknown to land at one's door in a box. There's a certain comfort in knowing that, mentally at least, you're properly prepared, whether for malabar pied hornbill, fairy bluebird or golden-fronted leaf-bird, that a decent attempt at feeding may be undertaken.

For years, the natural law, the one that said all stray birds must come to us, continued to have its way; birds appeared by one means or another and, from choice, necessity or general weakness of spirit, were included in the household. The enquiring phone call, the ring at the doorbell, the happening upon the miserable stray on the pavement were joined by a new danger, the pet shop.

Even now I have to be vigilant in pet shops, or at least in the ones that keep creatures in cages: small mammals, mice or hamsters or rats, a few cramped rabbits, possibly reptiles too, tanks of lizards, the odd small snake; the ones where, in a corner towards the back of the shop, there is a single large cage inhabited by one lone parrot, a sad-eyed African grey, a depressed-looking macaw. For me, the danger lies in the temptation to rescue. (Recently, I succumbed. I bought two white doves I did not need, since I have plenty of white doves of my own. On the third occasion that I was in the shop and they were still there, confined, unbought, I bought them.) Reason wrestles with inclination as I dart past beseeching eyes, past silent songs of siren bunnies, past the winsome, head-tilted, smiling charms of the lonely parrot who is there, I know, only because his previous owner bought him without having heard the astonishing, clangourous power of his voice.

The pet shop was how we came by Icarus. Bardie at the time was still young and Chicken had not yet arrived. Icarus was an eastern rosella, *Platycercus eximius*, like Bardie of Australian origin although of the Psittacidae, the 'true parrots', the other branch of the family, a small, compact-bodied bird of radiant beauty, red and blue and gold, of sweet and innocent face. He was unable to fly, which is why we were given him by the pet-shop lady, who no longer wanted the burden of continuing to look after a flightless bird nobody would buy. The feathers of his lower wings were tattered, some missing, lost to a psittacine ailment called 'French moult'. The name itself sounded oddly euphemistic. In those far-off, almost forgotten days before the Internet could provide all information, I searched in libraries, in bird books in the pet shops, but where I found mention of French moult, I found no reassurance. About French moult, it seemed, nothing could be done. About the causes, sources were vague. Only now I discover the probable nature of the affliction, which is viral. (I discover too the controversies, the problems of nomenclature, not unlike those of German measles and various social diseases, conditions and inclinations whose attributions are designed as insults to an entire nation. The British called the disease French moult, the French, British moult. Nearer to the truth, it seems, would have been the term 'Australian moult', for it appears to originate from there.) Whatever the case, the name 'French' stuck. It is the smallest comfort that, had I known all this, had I had, at that time, access to every piece of information revealed by the resources of the crowded ether, there would still have been nothing I could have done to restore Icarus's lost powers.

Because of his infirmity, we could give him no other name but that of the mythological unfortunate whose wings melted when he flew too close to the sun.

Icarus, the eternal terrestrial, was conveyed from place to place perched on the edge of a small, rectangular basket kept for the purpose, onto which he stepped with the airy nonchalance of the London businessman hailing a passing taxi. He was, we realised, already elderly when we got him, set into the patterns and routines of his bad-tempered ways. We put his house next to Bardie's. Bardie, still in verdant (or possibly more correctly argent) youth, was a source of continual annoyance to Icarus, who could be seen regularly standing on top of his house biting his feet in paroxysms of rage. For recreation, Icarus chewed his way through the very silly 'bird playground' (a structure with a tray, swings and ladders made – entirely unsuitably for parrots – from flimsy wood) that we had bought for Bardie, turning it in a very short time to matchsticks.

Icarus appeared to live pleasantly enough, to be busy and interested, or so we hoped, probably given more attention than he wanted, the sole threat to his safety being when, from time to time, in the process of waddling along the sideboard, he over-reached himself and fell from the edge, landing with the solid, unmistakable sound of parrot hitting ground.

It was many years later that we were given Marley, a South American sun conure. He had been bought in Palmers pet shop in London by friends in one of those impetuous moments of foolish, irresistible desire, the kind that override all caution. Looking at him, it was easy to understand why. He was beautiful, a small, brilliantly coloured parrot of yellow, green and orange. Before buying him, my friends hadn't heard his voice. They soon realised their mistake. On being asked if I would take him, I looked up a book: 'Sun conures,' it said with careful understatement, 'tend to be especially noisy members of the parrot family.' Marley's voice, like his beak, could have bisected wood and stone. It turned the air blue and shaky with waves of volume, the penetrating shrieks that were background to my life for a long time, his voice reaching me as I turned the corner towards the house, following me, inexorably, as I walked away. It may or may not have deterred anyone intent upon burglary.

Marley, when freed from his house, could work his way through the tops of the pine doors (as he attempted to do), through picture frames, mirror frames, tree trunks, beams, possibly girders. Only granite, perhaps, might have provided him with a moment's pause. When I was out, I worried from time to time that I might inadvertently have left his door open when I put him back after his bath, that I'd come home to find the house powdered: staircase, stairs, doors, lintels, furniture, everything systematically, enthusiastically destroyed by one determined, glorious yellow and orange bird.

I put Marley into the rat room, from where he could look out of the window at the garden, moving him to the sitting room, from where it was more difficult to hear him, when the noise became too much.

As with Icarus, we had no idea of his former life, although we suspected maltreatment, initiated perhaps as a response to the inescapable volume of his voice. His lack of trust, his wary confusions meant that he never became the easy, relaxed bird who forms a close, mutual relationship with a human. He was beautiful but not clever, stocky, square, with round black eyes rimmed with white. I showered him a couple of times a week, spraying him with a plant spray, being rewarded always by the look of unsurpassed bliss on what I liked to imagine was his smiling yellow face. He closed his eyes in ecstasy, dreaming of rainforests perhaps, of *florestas de varzéa*, the faraway world from which he sprang.

Adopting a parrot beyond infancy is always like adopting a child beyond first babyhood, one with an unknown but possibly unfortunate life to date. One adopts its past, the stamp of every owner – their voices, sounds, indications of mode of life – accompanying it. Its history is marked upon its personality, its view of the world, its optimism or otherwise, its capacity for happiness. As with an infant before it can speak, it can communicate its past only by gesture. There may be sundry wincings, flinchings, the parrot body language that hints at things unthinkable, the reluctance to allow the approach of humans, the abject terror at the sight of hands.

With the parrots we kept, there were never – like old lovers – names mentioned, those of former owners or persecutors, but there were the revealing sounds indicating time spent in unsuitable homes, the sounds of smoke alarms suggesting chips being cooked, things on fire, excessive smoking, or the incongruous wolf whistles uttered by the sweet-faced, staid, elderly Icarus. In a long-lasting example of cross-cultural transfer, of the sound dialogue that existed between them, Bardic learnt the wolf-whistle from Icarus and will still, years after Icarus's death, sound out with Icarus's voice, '*WHEE-WHYOO*,' repeated loud and often: '*WHEE-WHYOO!*'

In addition to other aspects of its mysterious past, a parrot may be of indeterminate age. The pet shop will not know, as the previous owner will not know. The bird may, for one reason or another, have been handed on from owner to owner over the course of a long and possibly distressing life. One day last summer, on a Sunday afternoon, Marley was in the rat room as usual, shouting, muttering, squawking as I passed to and fro chatting to him, cleaning, putting things into the washing machine. I was away from him for ten or fifteen minutes and when I went back there he was, as birds are in death, small, shrunken and closed-eyed, lying on the floor of his house, a few minutes dead. I don't know why he died. Nothing in his environment or food was different from usual. He was probably simply older than we had imagined. By then, he had lived here for seven or eight years.

Unaware of the convention by which one does not extend a hand towards a bird, the man who delivers coal, a small man of immense and day-enhancing cheer, invariably wiggled his blackened fingers towards Marley as he passed, poking them dangerously into Marley's house, shouting, 'Hello, my friend!', a gesture to which Marley, in violent and uncontrolled fury, responded by screaming loudly and hurling himself towards the fingers with intent to kill. Since Marley's astonishingly sudden death, the coalman says pensively on every visit, 'I miss my friend,' a sentiment which I appreciate deeply but, recalling Marley's voice, only partly share.

Birds had an interesting effect on us all. At some point after the purchase of Bardie ('purchase'! How the word demeans! How can one imagine, nineteen years on, that the vile medium of money could have been involved in this enduring, indissoluble, familial relationship?), Bec began to develop her theories of bird-rearing, the fundamentals being a sort of fusion between the ideals of Kropotkin, Montessori and a Code Napoléon for our time. Among its more notable clauses, a prohibition on the word 'cage' (the word 'house' being substituted), a de facto granting of full civil rights to all birds, which in practice meant never stopping them from doing anything that did not endanger their own well-being (ours being incidental), and enshrining in law the benefits of universal education, the necessity for perpetual intellectual stimulation and the freedom to avail themselves of anyone else's possessions.

When, years later, after she had left home, leaving me to try to maintain a certain mild discipline among the ranks (when a few birds would occasionally, for limited periods, be confined), Bec would arrive back and all would change. *Voilà! Le jour de gloire est arrivé!* As at the Bastille, closed bird-house doors were opened, all were set free. Everyone took part. Those who could fly, did. Those who could only wander along the sideboard, swearing loud, psittacine swears, did that. Anyone who was able and felt the need or desire flew to perch (uncatchably) on top of the pelmet board, twelve feet up, whence they'd shout remorselessly for hours. There was a tremendous, triumphant avian joy to the endeavours. I saw it. They had felt the air of freedom rushing through their wings.

We may have been the only people to reverse the usual beliefs regarding birds and cages. There were times, many times, when various birds, being expected to socialise more than they would have chosen, made a dash for home, settling with quiet relief onto their unassailable perches.

For years, under Bec's system Bardie ruled, or more accurately was allowed to rule. He flew shrieking up the stairs in Bec's wake, wandered freely (as he still does, more slowly, with greater trepidation) across the dinner table (as did Churchill's budgerigar, who would remove the silver implements from the salt urns before attacking his owner's cigar). He wandered freely over the books in the study, nibbling where he chose. The pages of many books in the house have, in addition to later damage inflicted by a magpie, neatly frilled edges, punched evenly along their length by a cockatiel.

a sun conure, small, brilliantly coloured, yellow, orange and green

Bardie would, while upstairs in Bec's room, on hearing the ringing of one of the set of Victorian bells outside the kitchen door with which I summoned everyone to meals, turn to Bec and say 'Come on!' peremptorily, flying to the top of her bedroom door before escorting her down the stairs for dinner.

In those far-off days of children, rats and assorted birds, we also had two canaries, given to us by an elderly man who came to the house to look at the rowing boat David had bought but never used, in response to an ad in the local paper. To reach the garden I had to lead him through the house, past bird-houses and birds, past calling and flying. He was, it seemed, not only a bird man but was cutting down on his own bird-house numbers. Did I want two canaries? I didn't but couldn't refuse. He did, I remember, spend a long time telling me about the more abstruse and complex intricacies of canary genetics. The boat remained unbought.

The canaries arrived a couple of days later and took up residence on a table in my study (on the grounds that it was the only remaining unoccupied space) in their solid, dull, pale-green-wood cage, a green that had the old look of arsenical paint but can't have been. All day they bounced ferociously on their balsawood perches, driving me to fury. They had qualities, but not ones I recognised easily. Perhaps if they'd been on their own, if I'd devoted more time and attention to them, been happy to settle for their song in another room, regarded them with more love and less irritation, they might have responded to Bec's theory of bird-rearing and proved to be more than the small singing pests that they were. I looked in the canary book we had hastily bought for indications of their possible lifespan. Females, it suggested, wearied by breeding, live five or six years. Males though, untrammelled by such demands, can live sixteen or more years. Both must have been elderly because in the course of time, not too much time, they went the way of all small singing pests.

The rats too, a constant presence for so long, thrived, became adults, aged, and one by one died. When the last one did, we took their well-used homes from the now-empty rat room to the skip at the recycling centre.

At some moment during the residency of the canaries we acquired Max, a starling. Aberdeen, since the discovery of oil, has become not only Oil City but Touchdown City, Move-on City; a migration stop, a place where oilmen from other places, from France, Holland, America, land, feed for the requisite length of time required for satiation before passing on to higher, or perhaps in the case of such subterranean activity, lower things, to the other places where oil is, or may be, to Houston, Baku, Luanda. Any day of any week, outside any house in this neighbourhood, a favourite of oilmen, pantechnicons may be seen. Large, square packages, taped and labelled for shipping, issue in the arms of burly house movers from the front door in a solemn procession to the van of the international moving company. Often, a laconic lady with baseball hat and clipboard will be seen sitting on the granite gatepost, counting. Neighbours, Louisianans and Texans, arrive and leave, our relationships ones that rarely develop beyond the rudimentary, the glancing, often briefly affectionate, mutually sympathetic, time-limited friendships between the nomadic and the settled. During these processes, lives are changed, those of children parting from schoolfriends; those of the rabbits and mice someone was persuaded to buy in an effort to establish more firmly the existence of the itinerant; the tank of stick insects – the bonds of love torn asunder. The dogs and cats, I assume, take wing and go too. The rest do not. (This morning early, I heard the last sounds from my neighbours' dogs, the ones I believe kept cats firmly away from the dove-house. For the past week they've howled in the garden in the wind and rain as removers have packed my neighbours' house in preparation for their return to North America. I was woken at four by a volley of muffled barks as they were being put into the car to be taken to the airport to be crated and, as my neighbour informed me yesterday evening as we said goodbye, herbally sedated for the flights taking them to their new life in Colorado.)

Max was an oil orphan, for who would take back to a country whose own, introduced, starling population has become a serious threat to indigenous birds, a single starling? The starlings of North America have unusual origins, springing, all 200 million of them, from the importation of a hundred birds from Europe in 1890 by one Eugene Schieffelin, an immigrant from Germany whose self-appointed task was to introduce to the United States every bird mentioned in the works of Shakespeare. The sole reference to starlings, the one that brought about this environmental tempest, is in a few not very interesting lines in Henry IV, spoken by Hotspur:

Nay, I'll have a starling shall be taught to speak

Nothing but 'Mortimer,' and give it him
To keep his anger still in motion.

The birds were shipped from Europe in two batches, one of sixty
and one of forty, and released in Central Park. Not all survived. The
ones that did were all too clearly successful. Their habit of stealing the
nest sites of indigenous, crevice-nesting birds, wrens, swallows and fly-
catchers, led to the introduction of laws prohibiting any further
importation of potential menaces. Writing of starlings in his poem
'The Great Scarf of Birds', John Updike hints at the scale of their success:

As if out of the Bible or science fiction,
a cloud appeared . . .

It was the lady in the pet shop where I bought bird seed, the one
who gave us Icarus, who asked one day if she might pass on our name
to an American customer who was leaving Scotland but couldn't take
the starling her children had found and reared. (I can imagine the
United States' response to the arrival of one to add to the 200 million,
even if he had survived the journey, crated and herbally sedated.) A
starling. What did I know of starlings? Nothing. It didn't stop me from
saying that, certainly, she could.

Pliny almost reassured me: 'the young Britannicus and Nero had a
starling and also nightingales that had been taught to speak Greek and
Latin, and, moreover, practised assiduously and spoke new words

every day in ever longer phrases . . .' Mozart too, I knew, once owned a starling, although his was bought for 34 kreuzer, from a shop in Vienna in 1784. By coincidence or chance, the starling was heard to sing bars of Mozart's piano concerto in G major, K453. Since he'd only recently completed it, it's not quite clear how the starling had learnt it, but it has been suggested that he had heard Mozart whistling it on a previous visit to the pet shop (such being the imitative powers of starlings). On transcribing the starling's song, Mozart wrote after it, '*Das war schön!*' The starling did, however, transpose one note, changing a G to G#, improving (or not) upon the great man's work. (A theory developed at length by animal behaviourists who have studied the amazing intricacies of the starling's ability to both mimic and sing, is that Mozart's later piece, K522, 'A Musical Joke', was based on the starling's song, an enduring tribute perhaps to the musical genius of one small bird.)

The American lady phoned. We discussed the starling, Max, when would be the best time to bring him, what he ate and what he liked. Towards the end of the conversation, the lady hesitated.

'There's just one thing I want you to know,' she said. Her voice was low, confiding. This was clearly a woman-to-woman thing. 'I thought you should know,' she said, 'he says the F-word.'

Not Greek and Latin? Had I had a moment's hesitation in agreeing to take him, that would have resolved me in his favour. I told David and the girls. We all waited breathlessly for the arrival of the swearing starling. *Sturnus vulgaris.* Suitable.

51

Max was brought, a glossy, gilded, bright-eyed bird. His house, though, had been grievously neglected. It smelt, a smell which I can still summon, sickening, sweetish with an undertone of mild putre-faction, emanating from the substance Superglued, if not for all time then for most of it, to the bars, the perches, the floor, the food dishes. We carried the house, and Max, to the bathroom, took the top, wire portion off the plastic base and let Max free. He darted enthusiastically round the bathroom, flapping his lovely angel's wings, alternately perching and flying while we immersed the noxious object in the bath to dissolve some of the noisome patina of staling shit.

The cleaned, respectable house, with its inhabitant, was placed next to Icarus on the sideboard, behind the dining table, in a row, Bardie next to the window, then Icarus, then Max. It was not unknown for the person sitting in front of his house to be the target of the well-directed projectile defecation at which starlings are proficient. Since it was Bec's habitual seat, and she was a well-known bird liberal, the useful notion of accepting the consequences of one's ideals was invoked.

Max was a bird of sparky charm, of enthusiasms and delights, his small, pointed green and gold feathers looking as if they'd been knit-ted by a craftsman's hand into a glistening, shining outfit of plain and purl. The prospect of cucumber enlivened Max's day, engaged his heart and, who knows, perhaps even his expansive starling's soul. He would hop, flap, beg and scream when a cucumber was produced from the fridge to be prepared, in a state of febrile anxiety until his portion was presented to him and jammed between the bars of his house, when he would set about it with meticulous care, spending a long time scooping out the centre, leaving a perfect tube, a translucent, jade-green tunnel of seedless flesh.

Max, it turned out, said lots of things, most of them unintelligible, but since one was listening for it, there it was. The F-word. It suited him, small, short-tempered, sassy Scottish street boy that he was. At dusk he would sing the complex, multi-layered, sweet evening song of the starling, bringing to me my Glasgow childhood of dark afternoons, twilight in a Scottish city, when phone wires were strung between poles, each, like a necklace, richly decorated with bead-like rows of glittering starlings. That was before the hand of sanitising bureaucracy skimmed the birds from wires, roofs, steeples, windowsills, removing the element of danger, the glorious, giggling frisson that enlived the progress of schoolgirls down Sauchiehall Street, a test of nimbleness as, in some complicated, recondite Highland reel, you skipped and side-stepped and pas-de-bas-ed round lampposts, through gutters, into doorways to dodge the steady but unpredictable precipitation from the massed starlings above. (Even now, I wonder how many Scottish city-centre buildings are held together by little more than an indestructible layer of starling excrement.)

Max lived with us for seven years, adjusting his song to the seasons, flying and shouting, and was old when he did what birds do, sank into himself, spirit shrinking, eyes clouding as he crouched in his small domain.

When his starling died, Mozart provided an elaborate funeral for him (compounding the widespread belief in his eccentricity, although the death of his father the same week may have inclined him towards greater attention to funerals). When Max died, we buried him quietly, with only the smallest ceremony, in the garden.

I think of him often, on the late-winter afternoons when I pass a derelict building in Langstane Place, a squarish granite building with boarded windows. From the darkness within the voices of starlings, probably hundreds, sound, their dusk song seeping, leaking joyously from every unseen aperture, a few last birds darting towards their home; and on the darkening afternoons when I watch starlings fly in a dusk cloud over the busiest part of town, while below people shop, oblivious. How starlings navigate during these astonishing evening displays is still not entirely understood, but they may take their markers from their immediate neighbours, each one in a crowd of thousands observing, turning and closing, moving with a fifteen-millisecond reaction, widening and spreading, drawing together in breathtaking, iron-filing flight. I used to watch from my vantage point at the window at work, overlooking the city-centre viaduct under which they roost, a starling city, an alternative world. Starlings organise themselves for the night in their social groupings, with adult males flying in to roost first, occupying the best places in the centre, whilst the young females, those last scatterings of flecks in the sky, sucked into the curve of the tunnels, have to make do with what's left.

'Just like life,' I used to say, probably tediously, to colleagues, watching. And so it is, just like life. (Although in the morning, the reverse is true – the old boys leave the roost first, to grab, no doubt, the best of breakfast, while the little girls wait their turn.)

In a recent cartoon feature in the *New Yorker* entitled 'A Guide to City Birds', Matthew Diffee provides the delightfully drawn profiles of five birds: a portrait and short curriculum vitae of snow goose, rock dove, sparrow, hawk and starling. The starling is Frank Scarpelli: 'Likes: Pizza. Dislikes: I hate freakin' cats and tourists and I've never been a big fan of birdbaths. I don't get the point, really. About me: I grew up in the Bronx. Friends call me Scraps. I work hard. Play hard. You gotta be tough in this city . . .'

4

Madame Chickeboumskaya

The years of keeping doves and other birds laid a foundation, encouraged a kind of enquiring acceptance of whatever, or whoever, might transpire or arrive, and so I was prepared. We all were. The evening the doorbell rang, we were ready, Chicken too, it seemed, smiling from her box, as infant rooks do, with their tragi-comic look, their corvid gravitas wholly at odds with the wide, frilled, amiable look of all small birds. This one, the offspring of the rooks that have lived for a long time in the woods near Crathes Castle, flying through and over its beautiful gardens, its yew hedges and its rose borders, peered from her box, her blue eyes interested, observing us as we were observing her. At once I was fascinated by her black, banded feet and legs, the fineness of her toenails, her pink skin erupting with dark feathers. The inside of her beak was bright, attention-grasping red, opening readily for food. Naturally, I took the advice of Kenton C. Lint. On the subject of feeding members of the crow family, his dietary recommendations were both reassuring and daunting: on the one hand, corvids would eat, it appeared, anything; on the other, part of their daily diet should include

Rodents: 40g

Chicks: 51g

Choice of insects: 14g grasshopper, locusts, crickets, beetles, grubs, moths or mealworms.

In feeding Chicken, I avoided the freshly killed or caught and gave her minced meat and eggs and chopped-up nuts instead. I had had no dealings with an infant corvid but from her learnt that healthy corvid chicks are vigorous, greedy, their beaks sturdy enough for a little finger, food-laden, to be thrust down the waiting throat to the accompaniment of the sounds of strangled gargling. She fed and slept and watched and I carried her around with me everywhere in her box. She sat beside my desk in daytime, on the kitchen floor as I cooked, beside the fire in the evenings. When I greeted her, she greeted me. I was entranced. After some weeks, she began to leap on to the side of her box to stand, clearly anticipating flight. Then, one day, she flew. I picked her up from the table where she had landed and put her back in her box. I realised that this could be only a temporary measure. We looked at one another, this small corvid and myself. *Well*, we seemed to say, at this moment of mutual, inter-species questioning, *what now?* What indeed.

By now it seems at best disingenuous to say that I didn't know enough of birds to consider reintroducing her to the wild. I wouldn't have known how to. Even now I'm not sure that I would know. Her home was fifteen miles away and there were no rooks nearby. The matter seemed simple. She had been brought to us and was, therefore, our responsibility.

We constructed a house for Chicken from wood and wire, fore-runner of her present abode, and placed it in the rat room, where the rats' houses once stood. Left alone for the brief periods she was, she began the first of her building projects, excavating the wall beside her house, picking determinedly at the plaster until she had removed the top layer. I didn't know then why she did it but there seemed to be no good reason to stop her. Holes can always be filled in.

I can't remember how much attention she got – less, certainly than she does now although she was always with us, always around us, playing with the toys we gave her, the rubber mice she liked to carry in her beak or to punish by shaking, pecking, bashing against the floor, for crimes unknown. She hid under tables, chairs, explored and began to take her place easily in the household. She was small, fluffy-feathered, and ever underfoot. We had to be careful not to stand on her as she pulled at the hems of our jeans or played with our bootlaces. She would fly on to the tops of cupboards and not know how to find her way down. We had to climb up to rescue her. She began to respond to each of us in an individual way, with a different voice, dif-ferent mannerisms, seeming to know from each of us what she might expect: a certain, limited degree of parental discipline from David and me, teasing fraternal playfulness from Bec and Han. Wisely enough, from the beginning, she understood that I was the one who fed her and although now I like to think that it was a bond of a different sort, I accept reluctantly that this might have provided the basis for our future relationship.

We progressed together, rook and human, and the knowledge, for the humans at least, was revelatory, mind-expanding, world-expanding. Chicken was clearly different from the other birds. I tried to examine the ways in which she was, to analyse what made her so. She seemed more inquisitive, more considered, as if her expectations of the world were broader. The doves' expectations and desires seemed confined – entirely reasonably – to the single-minded pursuit of the affairs of doves. Whilst the parrots too were intelligent and responsive to humans, they seemed simply to have a different world-view, one that extended less far than Chicken's. Chicken had an insatiable desire to find out. She wanted to know about the qualities of the small stones glittering with mica that she'd pick up in the garden, the purpose of the passing butterfly, what paper sounded like when it was torn. She wanted too to communicate, to be spoken to, to be heard.

Everywhere there were corvids and now I began to notice them, to appreciate them suddenly in another way. Driving, I'd see rooks as I had before, but with a new eye, a new acuity, the endless desultory pairs feeding, perhaps in the company of assorted crows, a few starlings, a handful of sparrows, in the grass of roadside verges, scattered as black flickers around every stand or thicket of trees, dusting, drifting, picking over the tilled surfaces of fields. They were all, as Chicken would grow in time to be, of sober mien, elegant of dress in well-tended black (except in summer when moulting renders them grey-edged and unkempt) with neat polished feet like tight, shining boots, somewhere between eighteenth-century Scottish minister (Henry Raeburn's 'Skating Minister' perhaps) and wealthy, black-clad, fashionable 1930s Parisian lady of distinguished years. I watched their walk, their gestures, what seemed to pass between them, in an infinity of behaviour I still had to learn.

There was more to know than I could have imagined; there was place and history and time. We knew corvids only in their wary, distant presence, in the sound of their voices. Corvids of one sort or another are found in many places in the north, as they are in most of Britain, among other birds, the lapwings and curlews whose calls are part of the sound of Scotland, the oystercatchers, gulls, herons, eagles, buzzards, hawks. Recently, I saw a map of rook distribution, which looked like a red scarf flung across the northern world. In urban settings, crows and jackdaws seem ubiquitous, pottering a few steps away, cautiously aware of us as we walk through the park, through the town gardens, black shapes in stark branches above us, silhouetted against pale clouds and sky. They're there on every stretch of roadway, every supermarket car park. Over farmland rooks fly, nest, feed and roost. One of the constants of many northern European cities is the presence of corvids; the crows I pass in a Warsaw park, the rooks on the spires and in the trees of Vilnius. In some cities, because of changes in farming practices and the greater availability of food in towns, their presence is fairly recent. Most corvid populations are settled, although rooks migrate from Russia, Greenland and Iceland south to Britain and other north European countries for winter. Worldwide, too, they're found in most places except Antarctica; in South America there are no crows, only jays and magpies. Populations differ in number and vulnerability, a few highly adapted species declining now to the point where they face extinction, the Flores, the Hawaiian and the Mariana crows among them, the latter two tree foragers, their habitats reduced by logging, as others' are by farming, industrial development or any of the other dangers humans introduce when their interests coincide with, and ultimately overwhelm, those of native bird populations.

Of the corvids found in Britain the ravens, *Corvus corax*, are the largest of all, with their neck ruffs of feathers and big, strong beaks. Rooks, *Corvus frugilegus*, have grey faces, long slender beaks and full leg feathers, whilst carrion crows, *Corvus corone*, are all black, neat-feathered, with shorter beaks than rooks. Hooded crows, *Corvus corone cornix*, look like crows wearing shaggy grey body-warmers. The jackdaws, *Corvus monedula*, have unmistakable silver eyes, short, pointed beaks and flattened panels of feathers at the sides of their heads. The other British corvids are not black: the magpie, *Pica pica*, is unmistakably, dazzlingly black and white; the chough, *Pyrrhocorax pyrrhocorax*, the rarest of the corvids, is red-legged and red-beaked. Jays, *Garrulus gladarius*, are colourful, light brown and blue and black and white.

Colonial nesters, rooks group in tight-knit, extended families, in rookeries of many nests, some containing many hundreds or thousands of birds. Some rookeries have been established for centuries, like ancestral homes with history behind them as rich and as long as any nobility. No coat of arms, no heraldry resonates as loudly, as profoundly as the sight and sound of rooks in their historic territory. In the vast rookery at Hatton Castle, a few miles to the north of here, there are thousands of nests where, every February, the rooks return to rebuild and repair their former homes. They choose to build nests in tall trees, in high, open situations, protected from predation by other birds probably by the proximity of other corvids and from the malign attention of humans by height. Rooks have occasionally adopted the dubious, unrookish and probably insecure practice of nesting on buildings as other birds do. E. M. Nicholson writes in *Birds and Men* that after the Napoleonic Wars, London rooks were seen to nest on the weathervanes on the turrets of the White Tower in the Tower of London and on the wings of the dragon on the vane of Bow Church. They can't have enjoyed the sophisticated life of inner-city London, even at these well-chosen sites, because the experiment was never seen to be repeated.

As with us all, human or bird, history has formed what corvids are, their behaviour the product of their long evolution, of lives often lived in close proximity to humans, subject to the demands made on us all to learn, adapt, survive. Their social organisation is complex, highly developed and whilst there are differences in the social lives of different corvid species, most are broadly similar. Ravens seem to live in the least social way, rooks and jackdaws the most.

The basis for most corvid existence is the monogamous pair. Many live in flocks, move in flocks, roost in flocks, separating to mate, nest and rear young. Most corvids live in 'nuclear' families, parents and off-spring, for the length of the breeding season, the raising and growing season at least, until the offspring are fledged. Some young will leave their parents, some will remain, sometimes for as long as a year, 'help-ing' to rear the next generation. Among the ones who leave, it has been shown that the females often travel further, putting more distance between themselves and their homes than males.

Certain niceties of interaction smooth the ways of crowded roosts, allowing corvids to live together without conflict in the numbers they do. There are necessary foundations to their relationships: mutual recognition, the ability to learn quickly, the skills of negotiation. To us they all look the same. Chicken (individual though she is to us) is in appearance as all rooks appear to be. I see many every day and whilst I've tried to see differences, apart from size or feather condition, a slight difference in the face, I can't. They do not, however, encounter such problems. Corvids' recognition of one another is a prerequisite for the kind of organised, highly social existence they lead, for recog-nising family members, accepting and reintroducing ones who have gone away and returned. They may even be rather better at both mutual and inter-species recognition than we are. Not only do they recognise one another, corvids can recognise individual humans, and there are countless stories of people involved in crow research of one sort or another being singled out from among large, busy crowds to be personally, individually subjected to harassment, a kind of revenge, no doubt, for what crows appear to regard as unwarranted scientific atten-tion. (Chicken certainly recognises many people, apart from the members of her immediate family. Some she greets with particular enthusiasm and what may or may not be expressions of welcome and pleasure.)

If corvid distribution is uneven, it's because each species has its preferences, or imperatives, dictated by the physical, evolutionary, climatic and social factors that have made it what it is. Ravens are birds of high, quiet places, of mountains; rooks, birds of farming country, in Scotland predominantly easterners; whilst 'hoodies', hooded crows, are the opposite. Choughs like rocky coasts, and jays woodlands. Jackdaws and magpies, like rooks, prefer the east. (Of all of them, the only ones that appear to like the north-west of Scotland are hooded crows and ravens. It may be that, in order to choose to live there, you simply have to like rain.)

From the first, I realised how little I had really observed the birds around me. It may be that the very ubiquity of corvids makes them all but invisible, beyond or beneath the interest of those who see them every day. Paradoxically, this may be why they appear to be noticed less, because they're just there, part of an accepted background, because their beauty is subtle, or unrecognised as such, their forms appearing at first glance to be clothed in unvarying blacks and greys, revealed only when close to as complex, shimmering, gilded with iridescent purple, blue, green. Their voices are perceived as harsh, unvarying, and except in rare cases, denying humanity the opportunity to hear reflections of themselves.

Perhaps, if the corvids we see around us were rare – had they, as others, already set their neat, black feet on the increasingly swift pathway to extinction – voices would be raised, and money, and campaigns set up, but since they are neither they require no such attention.

In time, Chicken developed her full adult plumage and became as she is now, beautiful, as are all crows, rooks, ravens, magpies. She is in every aspect, as they all are, in every movement, a sharp, tenebrous grace in her stillness, in her wings and feet and head. Corvids' beaks are balanced, proportionate, burnished and striated like the metal of a Damascene sword. The Japanese word '*shibui*' most encapsulates for me what they are and how they look, a word defined as 'austere, simple, quietly beautiful'. (It is no surprise to me that they are portrayed at their most exquisite in the art of Hiroshige, Hokusai and others, the art of a culture that sees crows so differently from our own.)

Because we seldom have the opportunity to be close enough to see the colour of their eyes, we may not know the depth and expression of the chestnut irises, the black pupils of the rook, the darker, fulvous plum and brown of crows and magpies.

Corvids, by being in the main black, are seen as representations of darkness, sources or conduits of evil, possibly messengers of dark forces. In fact, black feathers are protective, the strongest of all feathers, offering both camouflage and metabolic advantages over white feathers in their greater absorption of solar energy, which allows corvids to live more easily in the wide span of geographical territory they do, protected from the sun's rays and insulated against the Arctic cold.

As Chicken's feathers grew to thick, piled black, the irises of her eyes too changed from blue to grey-blue, then to deep, rich brown. She began to show the characteristic grey cere of the rook, a mysterious, ever-changing landscape. It's one of the aspects of her I could know only by closeness, by watching over time; that the grey portion of her face is not static. The texture of the skin, which reminds me of lizards (and makes me think of her distant dinosaur relatives), is of a strange and wondrous beauty, like lava or pumice, porous rock which erupts, melts back, is smooth then pocked, in an ever-altering pattern beneath the folds of grey skin under her eyes.

In summer, when she was small, we would take her into the garden. We began to clip her wings after the occasion when a sudden sound – I don't know what, a door opening, a voice from the next garden, a siren, another bird calling from a tree – sent her into panicked flight into a neighbour's garden. A child was dispatched over the wall and Chicken retrieved. Clipping involves the careful removal of secondary feathers, watching for their regrowth. When recently I omitted to do it, I realised only when, to my shock (and Chicken's), startled by something outside, she took off and flew round the high ceiling of the study in two stunned and fearful circuits. Her relief on landing was clear. The possibilities of danger for her, of becoming tangled in lights, colliding with walls, was too great. Again, I wielded the clippers.

wonderfully, eagerly enquiring

For the first years, when there were fewer cats in vicinity, we'd let her peck in the grass, investigate the flowerbeds under the bushes. Most of the time though she'd sit with us, on the back of the garden bench eating the aphids from the overhanging roses, their fine green legs waving helplessly from the sides of her beak, or 'sunning' – spreading her wings to the warmth and light (neither particularly abundant entities in north-east Scotland). The first time I saw her do it, I was transfixed by horror and panic – Chicken in sunshine, sitting on the edge of the garden bench, beak hanging open, head to the side, wings held wide and drooping (a posture we now call 'dying rook'), eyes veiled, apparently in the throes of a trance or coma. I didn't know what had befallen the unfortunate bird while my back was momentarily turned. A tentative calling of her name appeared to summon her back from this unknown realm, from her innocent pursuit of sunbathing.

No one knows the true purpose of birds' sunning. They may do it to help regulate their temperature, to increase their exposure to vitamin D, or to reduce feather parasites, but whatever it is, pleasure too appears to be involved. Since first seeing it, I notice birds everywhere spreading their wings in the sun, beaks gaping – blackbirds in the hedges of Union Street gardens, a thrush on the grass, the tiny robin on the garden table – all of them looking to the uninitiated as if they're in the last, painful throes of some alarming, rapid, fatal avian malady.

Adjusting to life with a rook was gradual, mutual, for us all, a process of interpretation, supposition, trial, learning the gestures of another's culture, the avoidance of the causes of fear or offence, matters of etiquette, slowly stepping one cautious step over the sacrosanct boundary into an unknown country.

Chicken seemed to enjoy being with us, perching under the table while we ate, hopping speculatively, carefully, on to someone's foot and, in time, their knee. We learned not to extend our hands too quickly towards her, or indeed towards any bird. Her wariness of hands, maintained until today, is entirely reasonable. One doesn't know what hands, or their owners, intend to do. But then, if Chicken is wary of fingers, we are equally so of beaks.

Corvid fears seem cultural, innate, rational as well as irrational. The Nobel Prize-winning naturalist Konrad Lorenz, in his book about animal behaviour *King Solomon's Ring*, describes his jackdaws' responses to seeing him holding a black, fluttering object (in his case, what he refers to as his 'bathing drawers') and being immediately surrounded by a crowd of angry jackdaws, trying to peck his hand, for they interpreted the object as one of their own, a dead jackdaw; other black objects, such as his camera, were regarded with equanimity. Chicken is used to most black objects by now, obliged to be perhaps by living in a household inhabited by inveterate and unregenerate wearers of black. She will though occasionally still complain loudly at the sight of a black dustbin bag.

The parrots we have kept, by comparison, have always seemed less afraid, more rational, less flighty in their fears, disliking cats and sparrow-hawks but regarding everything else with either calm or a degree of interest. It may be their different experience and history, being reared in aviaries, distant from their places of origin and from the circumstances of life in the wild that allows them a greater ease, but it may just be the way parrots are. (Bardie is afraid of chessboards. It may be that the bold pattern of black and white looks too much like the pattern of a snake's skin for relaxation in its presence but he may just dislike the game of chess.)

I don't know what she thought of ours, but we began to discover the grace of Chicken's demeanour. She made, and makes, me think of the fastidious conventions of courtly love, the way in which, with such refinement, she initiates and responds: each movement is careful. Not only rooks but even birds with reprobate reputations like magpies could shame humanity by their exquisite attention to manners, effusive displays of gratitude. Nothing, we discovered, is gracious like a corvid. Nothing displays such old-world, mannerly attention to others, such elaborate *politesse*, such greetings and such partings. Never, before meeting Chicken, could I have imagined the rituals, worthy of Japanese life at its most effulgently ritualistic, of coming in and going out, of waking in the morning and retiring at night, in acceptance and rejection, in speech and gesture, in meeting and making acquaintance, in the presentation of gifts; the bowings and callings, circlings and head-bendings, the solemn placing, as a morning gesture, of one cool black foot on the bare skin of my own.

At the beginning it had seemed simple, a question of responsibility. Then it started to seem less so, as I began to think of what it meant to keep a wild bird, one whose life in its natural setting would be so apparently alien, so dissimilar from our own. Dogs and cats are different. They have been bred for centuries for the lives they are to lead, for the small circuit, the delineated future. They live as they do, in a close relationship with humans, for the most part dependent on them because in the universe they have nowhere else to go. Birds, except for those bred in captivity, have plenty of other places to go. I am always aware that rooks are sociable, and that Chicken is without other rooks. But she isn't, as a member of the family, on her own. She's always with one or other of us, always within the sound of other birds, and if they're not the ones she might have expected, I hope it's consolation of a sort.

For years now, I've reflected on the facts of our coming by the wild birds we have, considered their prospects and their alternatives, what would have been their fate had they been anywhere but here. For each, Chicken and later Spike the magpie, the alternatives, I believe, would have been limited. Both birds were small when they first came here, too unfeathered for flight. Why they fell or were dislodged from their nests is impossible to know. Larger nestlings dislodge smaller ones while spreading their wings, practising for flight. Or, active creatures, they fall of their own accord. The trees from which they came are far too high for them to be restored to their nests by any practicable means. Reintroducing them to the wild might have been possible, but at best, in amateur hands, with individual birds, success is limited (everything I've read confirms it). Where would I have done it? How? Around us are urban gardens, busy streets, traffic.

Scattered by every roadside, the black corpses of corvids lie, feathers ruffled slowly by the wind. I don't regard their lives as cheap; the opposite, only infinitely fragile. The birds who have lived and live here have done so for much longer than they would have lived in the wild, although I am always aware of the ways in which they haven't lived, what they have been denied, either by my actions or by what might loosely be described as fate.

People have, I tell myself, for as far back as one knows, kept birds but there is no consolation in the telling because, for as far back as one knows people have similarly done things that are wrong. If I believe, or hope, that what I did was the best I could and can for them, I have to be ready always to answer to my sternest critic.

As we began to look at all corvids with new interest, we saw Chicken do as the corvids around us did. In time, we could recognise the complex series of movements of body, wings and feathers that told of mood and inclination. It may be the apparently sober colours, the lack of sexual dimorphism in corvids that obliges them to a subtlety of behaviour required less by birds that have more to show, more to flash, males with more brilliantly coloured feathers, sets of magnificent head plumes, vast, apparently bejewelled tails, elaborate songs with which to woo and win. We began to discern her state of mind from her stance, her walk, her feathers, to know that, when going about her day-to-day business, untroubled and busy, her head feathers would be smoothed to her skull, her auricular feathers (the panels of feathers by the sides of her head that cover the openings that are her ears) flattened, with no 'eyebrows' or 'ears' visible – the raised head feathers that indicate alterations of mood – no raised, irritated crown of Dennis the Menace feathers round the top of her head, a posture that indicates surprise, alarm, anger. Annoyance or some other stimulus, we saw, could bring this about instantly; when teased, or crossed in any way, she'd fluff her feathers, lower her head, adopt an aggressive stance, her leg feathers bagged out and full. When teased, as she often was (and still is) by Bec, she'd lower her head and spread her feathers, bow, fan her tail and lunge herself towards Bec in full, rookish fury. (Theirs, like Han's with Chicken, has always remained what appears to be a sibling relationship.)

'Crow!' Bec will say to Chicken, who becomes angry, begins to strut and spread her feathers. 'CROW!' This is clearly offensive, a crime against taxonomy. In our midst, a shape-shifter, a smallish, smooth-feathered, glossy rook one moment, a strutting, baggy-feathered, almost large, self-important, angry one the next. When, years later, we had an opportunity to observe a magpie, we'd know that the appearance of 'ears' on Spike's perfect, shining black head was a warning signal. *I'm getting angry*, it said, in preparation for violence and attack, the controlled fury of the probably dangerous; *why did you have to annoy me?* Get ready for the vengeance of one small and angry magpie, it said. One did. Their raised 'ears' always remind me of Batman's ears and are perhaps similarly indicative of righteous indignation or the highest moral intent.

We'd watch amazed as Chicken did what Bernd Heinrich (in his book *Ravens in Winter*) describes in ravens as 'jumping jack', when birds will leap up and down, hopping and calling, behaviour that occurs in the presence of an unfamiliar food source, a carcass perhaps, in a process of investigation. She does it still, leaping into the air, both feet off the ground, wings wide and flapping, tail raised, uttering quick, high yelps, 'Wup! Wup! Wup!', continuing for a half a minute before she stops and stands, slightly out of breath. In the absence of carcasses it's difficult to know why, what inspires her, what initiates it, and although I have tried to see a pattern or a cause, I never have.

I began to read books about corvids and to appreciate that what is known of their lives, the patterns of their behaviour and social organisation, has most often been learned by careful, painstaking and sometimes dangerous research, undertaken among the most observant, the most wary, quick and communicative of birds. I often read the accounts with awe, admiration and gratitude, knowing that if people who carry out research into corvids and other bird behaviour didn't climb trees, the very high ones that are the first choice of most corvids, didn't spend long, freezing winters in cold, wild places, watching, tagging, measuring, didn't set up experiments that, because of corvid wariness, fail after lengthy preparation, we would know considerably less than we do and people like me would not be able to exclaim from their armchairs in revelatory recognition at a piece of corvid (or other avian) behaviour they read about in one of these astonishingly detailed accounts. Bernd Heinrich describes climbing high trees, precariously, dangerously, in storms in order to reach a raven's nest; he writes too of those who have sustained injuries while carrying out research, of Thomas Grunkorn who fell out of a tree and broke his back, of Gustav Kramer who was killed falling from a cliff while studying wild pigeons.

In their book *In the Company of Crows and Ravens*, researchers John Marzluff and Tony Angell describe being picked out from among forty thousand others on the campus of the University of Washington by crows who, while happily walking among other people, fly away from them. Kevin McGowan of Cornell University was routinely identified among crowds, followed and shouted at by crows he had studied (and some he had not). Bernd Heinrich, attempting to discover what it is that allows ravens to identify each other, describes experiments with his own ravens, who happily accept him but no one else. Using a variety of techniques – swapping clothes with other people before approaching the ravens, changing his outfits or elements of his outfits, wearing masks, wigs, sunglasses, making grotesque faces, limping, hopping, carrying a broom – he proves that if it's him they're not fooled for long by any subterfuge, that, just as the visual clues the birds use in recognition of humans are diverse, so probably are those they use in identification of other birds. (Included in the account is the unforgettable sentence 'After my thirteenth approach in the kimono, they again allowed me to get next to them.')

Interestingly, the broom was the one thing the ravens never accepted, as Chicken will not. Even after years of close, daily acquaintance, Chicken still runs away at the sight, hides under the table until it's put away. Corvids must know something I don't about brooms.

Gradually I began to realise how much more there was to know, that what I had learnt, what I had observed from Chicken, was just a beginning. I read in the introduction to Kenton C. Lint's feeding instruction for corvids that crows and ravens can live for twenty to twenty-five years in well-planted aviaries. While this isn't a well-planted aviary, I wondered if it might do instead. It seemed that with good fortune, or whatever else it might take to look after this bird properly, we could well have some time to spend together. I hoped it would be so.

Part II

LIFE WITH CORVIDS

becoming one of us

5

The Hole in the Kitchen Wall

During the years of sharing this house with birds, the kitchen and rat room have always been communal space, inhabited democratically, equally, by both human and bird. The study, on the other hand, is the *salon privé*, a place where no flying is allowed, no high-level depredation, no intrusion by the unwarrantedly curious or wilfully destructive.

It's a long time now since Chicken moved into this room, and although we work in it together, our status in it is different. For Chicken, the room is home, where her house is, while I am a mere sojourner, spending time in it during working hours and sometimes in the evening, when I prevail upon her generosity to share with me a room that, although clearly not mine, is known erroneously as my study. It's a room with large windows and a French door which opens on to the garden, on to bushes and stones and bird feeders, a stretch of grass beyond, the dove-house, a small and overgrown pond. Outside the window, the garden birds and doves are perpetually busy, doing what they have to do, while inside Chicken and I occupy our time as we do on most days, me writing, she pottering earnestly with deep, driven intent, hiding things, perching on her branch in contemplation, eating, bathing; throughout the year, carrying out the many demanding imperatives of each season – in spring, preparing for nesting, in summer, dealing with the uncomfortable time of moulting.

Chicken's house, while beloved to her, is not beautiful. It's a shanty house, a dwelling more suited to a *favela* than an otherwise elegant, high-ceilinged, corniced room. The favela house was constructed hastily by David years ago, when Chicken was moved from the rat room at the back of the house to this more central, sociable situation, an arrangement she appeared to welcome.

The house, which stands on a worn, holed, raggy Persian rug of pink and dark blue, has a wooden back and door and sides of galvanised garden mesh. A metre high, a metre and a half long, it has stout, removable and, more importantly, washable plastic trays for the floor, which I cover every day with fresh newspaper. It has a wire door with hinges of wire and string. The door is rarely closed but if it is, during the cleaning of the floor perhaps, it is swiftly reopened thereafter by Chicken who pushes it triumphantly with her beak. She has a set of feeding dishes of white plastic with hooks, which I can attach to the wire of her wall. She can, equally, unattach them when, in response to whim, she decides that she would prefer her meal to be eaten from the floor. In spring, when her behaviour changes, she likes to carry the empty dishes in a clattering procession through the house. Often, it's the first sign, the first small intimation, even on days in mid-February when fierce wind blows a few flakes of reluctant snow or on the grey, pouring days of early March, that beyond one's own narrow perceptions of prolonged and dreary winters, others have already begun to scent and sense that the season is about to change.

Two perches made from stout branches from the garden are secured on different levels, at different angles, on to small wooden brackets and held with screws. A passerine, a perching bird, her feet close as her ankles bend, the tension of the tendons curling her toes, which hold and tighten to keep her safe.

On the floor are two large stones upon which Chicken likes to stand, as the stag in a Landseer painting, nobly, next to the water dish of heavy stoneware, which is also her bath.

Hanging from the wire of the roof of the favela house there is a series of bells of different sizes and types which Chicken rings with the boundless, energetic enthusiasm of the church bell ringer, except that, in her case, protest rather than campanological fervour is her motivation, an expression of outrage or disgust, at the sound of music she doesn't like, perhaps, or at the regular appearance of cleaning equipment, to which she holds stern, enduring objections.

Chicken and I are both used to the favela house and although David often offers to improve it or to renew it completely, on Chicken's behalf I decline. She's not concerned with home improvements. She loathes change, reacts with terror, running to hide in the kitchen or behind the sofa in a torrent of loose-bowelled shouting, when one of her branches, un-notched from its moorings, falls down. She notices the smallest of changes to her environment and reacts with suspicion, fear and scowling resentment. When, a few months ago, I bought a cupboard for my papers to install in the study, Chicken was shut in the kitchen while furniture was shifted from place to place. There was, of course, no thought of moving the favela house. The room was reorganised around it, table moved from one side to the other, cupboard installed, lamps and pictures replaced, and when the room was satisfactorily restored to order, the door opened to allow Chicken to return. Her caution was total. She peered round the corner, hopped back, peered in again, a portion of beak, part of a small grey face appearing, retreating, reappearing, retreating again in the minute process of establishing that her house at least was untouched, in every manner and respect the same as when she left it. She looked with suspicion and probably contempt at our attempts at interior decor but, since her own establishment was intact, she was prepared to live and let live.

While the favela house lacks style, it's in this modest dwelling that Chicken has place and opportunity to do at least some of what all corvids do: preen and perch, roost, hide and retrieve, call and feed and bathe.

In spite of our years together, aspects of Chicken are difficult to know. Her sight, being markedly different from ours, is one. Birds are thought to have the best vision of any vertebrate, the placing of their eyes determining not only nature and scope, but their capacity, depending on their species, to watch for predators or prey. Raptors and owls, both predators whose eyes are at the front of their heads, have binocular vision as we do, whilst birds such as doves, whose eyes are on the sides of their heads and are more often prey, have monocular vision, the ability to see objects with only one eye at a time. Birds with monocular vision usually have a very wide visual field; some, such as the woodcock, whose eyes are placed far back on its head, can see what is behind it more accurately than what is in front. ('It could better tell where it has been than where it is going,' according to the distinguished political scientist and nature writer Louis J. Halle.) Whilst human vision is limited by having three light receptors, or cones, by which colour is seen, birds have four or five spectrally distinct cones, sensitive to a far greater range of light waves, including ultraviolet, than humans, which may allow them to see twice as many colours as we do. They can also distinguish between wavelengths of the ultraviolet spectrum, a visual acuity that gives them fine perception in foraging, letting them identify a wide range of colours of food and natural objects. Raptors, although the most visually acute of birds, are thought to be able to see fewer colours than passerines.

Chicken's sight, once alarmingly acute, the sense which in birds is most highly developed, must by virtue of her useless eye be lessened, although most of the time it doesn't seem so. (David has given his opinion on the neurosurgical, or at least neurological aspects of her eye problem, which is possibly a cataract. His sister Zanna, an ophthalmic surgeon, has been consulted on the ocular. Both have agreed. The combined weight of their knowledge suggests that they're not entirely sure what it is and in any case there's nothing to be done.)

Chicken's right eye appears to compensate for the deficit in her left, for she is still able to see very small objects, a pine nut I have given to her lying on the floor, a crumb of something she considers desirable under the fridge. Because of the limited movement of the eyes in their sockets, birds' necks have greater flexibility. Chicken has the enviable attribute of being able to turn her head upside down to look underneath the fridge or sofa. When we're sitting together as evening wears on, she'll descend to a roosting position, her feathers spreading a warmth across my knee, and she will appear, on one side, to be asleep. On the other, her eye is wide and bright and totally awake. 'Is she asleep on your side?' we'll ask. The hemispheres of a bird's brain alternate in waking and sleeping, the eye on the side of the somnolent, slow-brain-wave hemisphere closing. They can also sleep with both hemispheres at once. Research among ducks has shown that the ones at each end of a row of sleepers will keep the appropriate eye open, watching.

Chicken trips sometimes over an item left lying on the floor, a piece of paper, something small, and I don't know if it's age or sight that causes her to do it. My heart stops briefly when she does, for whatever this is it is quite new.

For reasons unknown, she nurtures a residual hatred for my specs, as a result of which I operate most of the time (as I suppose she does, on one side at least) in a hazed fog caused by her habit of standing on my shoulder and inserting her beak behind my ear to tweak my specs off by one leg before sending them skimming, with the sound of lens scraping wood, across the floor. The fog lifts only if I have time to replace my scratched, semi-opaque glasses when, for a short while, I'm astonished anew by the fine-etched clarity of the world.

Almost equal to her dislike of my specs is her dislike of cleaning or electrical equipment, usually because of a dangerous combination of the sight and sound. She is scared of vacuum cleaners. In common with many people, she hates computers, their blue-white stares, their sudden vulgar, explosive bursts into colour, their tendency at certain moments to talk. She has accustomed herself only reluctantly to mine, grunting with mistrust and displeasure at the sight of the screensaver (sentimental images of the natural world, forests in mist, ladybirds, dew-laden plants, chosen with her in mind, since she disliked even more the alternating planets on my previous screen). She dislikes too printers, cameras, laptops, music-making machines of every sort, the television (which is not kept in the same room as her and is, in any case, never watched): their winking, malevolent green eyes, their watchfulness, the inscrutable nature of their intentions. She was distressed, it was clear (as I was), by the single terrible scream, the loud and dangerous whirring and gusting, as the logic board of my quite new computer succumbed one Saturday evening in March to the equivalent of a ruptured brain aneurysm, and died.

simply for pleasure

Perhaps she knows, as the rest of us suspect, that of all of this no good can come.

Chicken does not like to be ignored. She'll pull insistently at the legs of my jeans as I cook or iron. She'll try to knock the book from my hand if I don't pay her attention, pecking at my sleeve or elbow to invite me to talk. Often she'll burst, like the alien in *Alien*, through my newspaper, leaping on to my knee as I sit trying peaceably to read.

It has been suggested that there are differences in the way the various species experience time, and that, for those having faster time-scales by virtue of their shorter life-spans, the perception of time will be slower, that one minute of their lives will be as several minutes of our own. I don't know how Chicken experiences time; if we pass a day, an hour together, we do so at the same pace.

I know little of her sense of smell, although in most birds it's generally regarded as poorly developed (an exception being pigeons, in whom it may play a significant part in direction-finding, and some other species for which extra olfactory sensitivity is required for navigation or safety). I don't know if it's smell or sound that draws her instantaneously to the kitchen when butter, her favourite food, is removed from the fridge. I've tried to test her, removing it as silently as I can, but my silence may not be hers.

About her sense of taste, I can judge only from observation. Although birds have fewer taste buds than humans do, they have distinct preferences. Chicken can differentiate between foods of vaguely similar texture, a banana or an avocado, rejecting the former, accepting – with delight – the latter. The fat content of foods appears to be of importance, although by now she is able to distinguish between different types of cheese, between Bel Paese and aged Gouda, which she does like, and Camembert, which she does not. Recently I offered her a piece of a good Mull cheddar which, with the air of the dedicated oenophile contemplating a bottle of fine wine, she examined for a moment, appearing to sniff it. After consideration, she picked it with her beak from my hand and hurled it to the floor. This was the final judgement. It was not even worth hiding under the carpet. As most birds appear to do, she loves egg yolk in any form. Unlike Jakob, the pet raven about whom Bernd Heinrich writes, she rarely has the opportunity to sample Chinese food but would, I'm sure, try with great eagerness the Hessian cheesecake beloved of Jakob. Unlike him, though, she rejects most fruit except cherries and even then I wonder if it's the resultant sanguinary mess created by a rook with a red fruit that is the attraction. If food is to her taste, and movable, she'll hide it. If it's hard and will be softened by dipping into water, she'll dip it into water. As a result, in the morning, her water dish floats with portions of breakfast cereal and bread, which I try to fish out before morning bathing begins.

At mealtimes, Chicken likes to participate, either pottering on the floor or, given the chance, standing on someone's knee. She gazes calmly across the table, as a guest at a feast, and appears to expect at least to be recognised as part of the company. On Friday evenings, she recognises, as all the other birds who have lived here do and did, the sound of Kiddush, the lighting of candles, the recitation of blessings (my one enduring nod towards the life spiritual), and will, as Bardie does and Spike did, express eager, vocal anticipation of the coming of Shabbat or possibly of the cutting of the *challah*, the home-baked egg loaf that they, naturally, will share. Such *frumers*! Who'd have imagined!

I have only once committed the crime decried by Lord Byron, another corvid owner, of standing on Chicken's foot. (How it's happened only once, I don't know. She likes to follow me, to stand behind me when I'm cooking, to place herself just where my foot will land when I step off a chair if I've been putting something away in a high cupboard. She is invisible in the dark.)

Lord Byron, best known of course for other matters entirely, seems to have been a man of catholic tastes in both people and animals. According to a letter written by his friend Shelley during a visit to him in Ravenna, his house was shared by dogs, monkeys, cats, peacocks, guinea hens, an eagle, a falcon, an Egyptian crane and a crow.

In his diary of 5 January 1821, Byron writes of feeding the hawk and the 'tame (but *not tamed*) crow'. On the 6th, he complains of his crow being lame: 'some fool trod on his toe I suppose . . .'

I pass a statue of Byron every day. For a brief time in 1794, he attended Aberdeen Grammar School when he was living here with his mother. Despite the brevity of his sojourn, his statue stands in bronze-robed solemnity and magnificence in front of the school's splendid granite façade. As I pass, I salute the man. I am unmoved by Lady Caroline Lamb's famously damning designation of him, because nothing can alter the fact that it speaks well of a man when he cares about his pet crow's toe.

In the past few years, Chicken has become reluctant to go outside. Now, in summer, I open the French windows and she potters to the door. I have a small enclosure of wire to allow her to step outside but not to go too far. There are cats next door and no children to rescue a fast and frightened rook. But then I am more frightened too. Perhaps, for both bird and human, fear creeps up, becomes the adjunct of age, when the days of immortality grow dimmed, tarnished over by the knowledge of death, the sight of cats peering in windows, the passing sparrow-hawk in the sky, the adjunct of our own knowledge of all things threatening. On a late-November afternoon, I see a hawk flying against a cold, silvered sky, the half flap, half smooth glide, the silhouette that can reduce a safe, protected indoor bird to shrieking terror.

At nightfall, when it's particularly cold, if there has been snow perhaps or a thick frost is coating the branches and the grass, I try to close the heavy curtains in the study to keep in the warmth. The result is inevitable. Chicken runs from the room, scared and outraged, into the kitchen. She refuses to return, although it may be roosting time, until the curtains are open. I have to coax her, or even chase her back into the study so that she can witness the extent of my compliance with her wishes. I think of the warmth, this small representation of the sum total of world's resources being depleted through the windows and French door of the room but none the less I open the curtains and tie them back. Chicken returns, hops onto her branch, runs the edges of her beak angrily against it, still with a look of suspicion and, I think, the faintest tinge of resentment at the power that my ability to remove, on whim, her peace of mind represents. She adopts a confrontational pose, bows and caws. I look out at the snow-lit garden, at the black branches and weighted, moleskin sky. If I'm honest, I prefer it this way too. I too like to see out, to see the sky, and I understand her need to see it, to see stars if there are any, the moon, the tops of the branches of the *Viburnum carlesi* flowering in midwinter opposite the window, the shut door of the dove-house. Occasionally, with luck and the correct atmospheric conditions, we will see the light of the aurora borealis, even in the middle of an over-lit city, as it throws its luminescent canopy of glowing, shimmering chiffon to fall in green and pink and gold over the city and the garden.

Often, as I sit at my desk, the sounds behind me are of tearing newspaper, of beak and claw as Chicken carries out some elaborate ritual of domestic rearrangement. It may be early morning, and Chicken still at breakfast, the consummate corvid process of eating and hiding, the sound of her beak closing on fragments of bread and butter alternating with the rustle and rip of the paper that carpets the floor of her house, as she hides things. I don't know her criteria for choosing what's to be hidden, whether, as a child might, she's keeping the best bit until last, or if she's stowing the dull, butterless portions to be dealt with when the rest is gone, or when the food runs out, an unlikely enough occurrence in this household. Since I am the only person with her, I assume that she is hiding it from me.

I don't take it personally. All corvids hide, or 'cache', things. Caching, the collection and storing of food, is behaviour much studied as an indicator of a range of avian abilities extending far beyond the simple act of storing food for winter. Many birds cache, among them tits, woodpeckers and some raptors, but it is the corvids in which the behaviour seems to be at its most developed and sophisticated. If I hadn't watched Chicken over years, I wouldn't have known this. I would have known from reading that corvids cache but I wouldn't have understood the obsessive, busy nature of caching, an activity that, for Chicken, is natural behaviour elevated to the status of art.

Although some authorities on the subject suggest that birds who have food available to them at all times don't cache, Chicken does. It seems as much an occupation, a profession, as an insurance against hunger. She chooses her cache site, carries her bread or other food to it and begins the long process of concealment. She has many cache sites – under rugs, under the newspaper in her house, inside the large velvet floor cushion in the study whose seams for years she has been painstakingly unpicking to allow her access to hide things (I no longer like to think about exactly what). After the initial hiding, she'll tap down the paper or rug under which her stash is hidden, bending fussily, tidying, smoothing, standing back to make sure it's fully concealed before turning the paper or rug over, picking it out and beginning again. She hides things not only in her house and under her own rug, but also under the Chinese rug upon which the dining table stands where, inevitably, I or someone else will unknowingly stand on them. Later, I will take the paint scraper – the tool most useful for the person who cleans up after a bird – and remove the flattened, unidentifiable mess.

When it comes to caching, nowhere, or rather no one, is sacred. Chicken is on the floor. I am eating lunch. I give her a flake of poached salmon from my plate. She takes it with alacrity and immediately begins to cache it. Her choice of site, were I a habitual cacher, would not be my own. Thrusting her beak under the hem of my jeans, she wedges the fish between the laces of my boots. I do nothing, for it will not, I know, be there for long. She hovers a moment then retrieves it. She carries it into the hall and, since I haven't yet noticed the odour of rotting salmon, I have to assume that it was eaten. I search the turn-ups on my jeans regularly to make sure that, if Chicken hasn't removed an item recently cached, I do. When a food finds special favour she takes particular trouble over the caching, wrapping the item first in whatever paper is to hand then unwrapping it and wrapping it again. Often, I see one of the long curtains in the kitchen moving, apparently of its own accord, as Chicken stows her treasure and retrieves it again. Her current favourite for caching is goat cheese. One piece can occupy a day as it disappears and reappears, is nibbled and hidden again, and ultimately unearthed (by me), staling under the rug beneath her house.

Perhaps, though, the greatest monument to her art, the most supreme of all caching sites, is at the foot of one of the kitchen walls, an area above the skirting board where both paper and plaster are missing. The excavation – for it is more excavation than mere hole – exposes four horizontal lines of lathe. Just visible in the cobwebbed darkness behind them is a grey hint of plaster, mortar, horsehair, the materials Victorian builders used to fill the space between the solid granite slabs. Five or so inches by about eight, with the gently rounded curve of a graph rising, then falling, the area of the excavation expands minutely but inexorably, day by day, year by year, with the slow, eternal movement of the tectonic plate. Chicken is the architect, the archaeologist of the project. I no longer attempt to fill it in or to prevent its expansion because I have realised that to do so would be to thwart her instincts and, who knows, perhaps her ambitions too. What her ambitions are, I do not know. I accept though that the excavation is, if not her life's work, then a significant part of it, inseparable from her profound corvid need and desire to hide things and then find them again.

Into this space in the kitchen wall she continues her unending work of posting and collecting. She enjoys the hiding of both food scraps and paper in her excavation. A particular pleasure is in removing from my wooden trug of magazines, newspapers, assorted letters and things-about-which-something-must-one-day-be-done, a slip of paper – any paper, an old receipt, the envelope from a bank statement – which she then posts between the slats of the lathes before redeeming it for future use. (I assume that nothing of too great importance can have disappeared between the lathes in the years she has been busy with her filing, because no repercussions, final demands or angry insistences for immediate attention have occurred.)

I don't mind the hole in the wall. Indeed, when recently, the kitchen was being redecorated, the principled decision was made not to disturb or alter it. It has its own aesthetic: not quite shabby-chic, more a bold counterpoint to the elaborate *trompe-l'oeil* you see from time to time in interior-design magazines, the grandiose schemes, Greek temples or improbable pastoral idylls, marble blocks, Etruscan arches. I have on occasion been tempted to paint a *faux* mouse peering from a corner, but haven't because I feel that this would demean the project and might, who knows, earn Chicken's lasting contempt for my frivolity.

There is a way in which I think of the kitchen wall as more than just part of the pattern of her life. It belongs to her. Into this void, some things fall beyond even the grasp of her long beak and tumble, I suppose, into the space below the house, where they are either removed by the odd mouse to be shredded for bedding or else continue to lie, awaiting the day when the fabric of the house, like the fabric of all things, will founder, fall into desuetude, the day when archaeologists, not knowing why paper might have been stored in the foundations of houses, will come inevitably to wrong conclusions. I reflect on the sadness of their loss, that they will be unable to wonder appropriately, to document and laud as they should, the indomitable qualities, the charm, the intelligence of corvids.

6

The Black Airts

It didn't take long for us to realise that our love of corvids was not universal. The girls' friends in particular regarded us as an outpost of the Addams family, intriguing, strange, potentially sinister. The only grounds for their view (as far as I know) was the presence of Chicken.

By now, I'm not even surprised by people's reactions to sight or mention of her. I try to explain. 'Like a very small black Labrador,' I've heard myself say, in an infinitely feeble attempt to find a way to make the matter comprehensible to those who assume that the rook I have is one who visits the garden from time to time, touches down, feeds, moves on. 'She lives in the house,' I say, trying to find a way to bridge the unbridgeable, the notion of 'dog' being the one most useful to invoke, that idea of comforting canine solidity with which most people in this society at least are familiar, yet which seems so at variance with the comparative smallness and fragility of birds, with the alien concept of 'bird' for those who cannot imagine even being in close proximity to, or the experience or sensations involved in spending time with, a creature normally seen as wild. It's not strange to share one's house with a furred quadruped, but it is to share it with a feathered biped. I recount some of Chicken's qualities in the face of their disbelieving gaze. Can anyone really accept that a rook might be companionable, intelligent, charming? With difficulty.

I still don't know what to say of – as I didn't know at the time what to say to – the person, friend of a friend, who recently, when I talked of crows, said, 'Crows? Horrible birds!' I'm still undecided about which shocked me more, her tactlessness or the bleakness of the inner world her words conjured for me. She's a teacher of English. Don't the frequent, resonant allusions to corvids in art, poetry, literature impel her to something better?

If others think it, few have said it. If they think it, I tell myself it's because they don't know. How can they know? But then I wonder why they don't. These birds live among us, above us, beside us. How can we know so little, or nothing, of them?

It has been, for the long duration of Chicken's presence in this house, the commonplace for every person who first sees her to ask (with a certain insulting casualness in the use of the pronoun), 'What is it?', the question demonstrating only that many people cannot distinguish between one corvid and another, between the passing carrion crow picking in the urban park, the hooded crow, the sharp-beaked, silver-eyed jackdaw, the baggy-trousered rook. Is there any reason they should know? Once I wouldn't have known either.

'Is that a raven?' they'll say on seeing Chicken standing on top of her house or perching on her branch. I suspect they don't really think she's a raven because they don't know what a raven looks like either, that it's a name, a word risen from that part of the ether in which is kept a small list of the names of black birds. (They probably won't ever have seen ravens, except perhaps during a visit to the Tower of London.) What they want to know is which kind of crow she is. I tell them. I omit to say that, by contrast, she knows very well what they are.

Often, the people who ask are the same ones who, by worthy, rapt and enthusiastic attention to wildlife programmes on television, know the minutest workings of the inner lives of polar bears, of anteaters, hummingbirds, frogs. They've seen unfurled before their eyes the most intimate transactions in the lives of other creatures, wooing, mating, birth, all in magnificent colour and irresistible detail, each undreamt-of habit, each hitherto opaque, obscure aspect of nesting or feeding or defecation, but will say, 'What is it?' when confronted with the one bird they see every day, making me reflect yet again on the oddness of humanity, which, in its desires and its yearnings, wishes to find life on other planets, other civilisations, but knows so little of the civilisations around it. Implicit too in the question about what Chicken is, I realise, is the unspoken word 'why?'

It's not only that people don't know what Chicken *is*, they don't know what she does. The question 'What is it?' is prompted not by curiosity alone, but also by fear. People are scared of Chicken, unlikely people: huge men see her and instantaneously a shadow of anxiety alters their faces.

'What is it?' they say, hoping only that I'll take it away. Lads only a generation or two from Aberdeenshire farming life hover nervously until, on shutting the door to the study or kitchen, danger is past. What is it? A rook, my boy, a rook, a bird of the kind by which, every day and in every place, on every roadside verge, overhead in every tree, you are surrounded. A rook, the like of which your farmer grandfather in Strathdon or the Mearns will have waged daily (if unnecessary and futile) battle. I want to ask them what they're afraid of but don't. I was frightened of birds, at the beginning, not simply ignorant. I remind myself that I was afraid not only of corvids but of doves too, of all birds, for I shared what now appears to me to be this near-universal apprehension, one that lies in not knowing what birds may do or wish to do, an unfamiliarity with their habits, their ability suddenly, terrifyingly, to fly. The history is too long, the fears and superstitions too deep-rooted for flippant questions.

The name of James I of Scotland is one we seldom mention in Chicken's presence, because, for all the worthy civic efforts he undertook on his return to Scotland in 1424 after eighteen years of captivity in England (his visionary enterprise in rebuilding the palace at Linlithgow, recently destroyed by fire, his enthusiastic, new-broom frenzy of legislation relating to governance, law, the ownership of mineral rights, and the imposition of restrictions on playing football), among the many laws he passed in 1424 was one decreeing that rooks, for their alleged damage to cornfields, should be killed in their nests, any farmer being found with nesting rooks at Beltane being obliged to give up the relevant trees to the king, until payment of a fine:

> Item, for they that men considderes that Ruikes biggard in Kirk Yardes, Orchardes and Trees, dois great skaith upon Cornes; It is ordained that they that sik Trees perteinis to lette them to big & suffer on na wise that their birdes flie away . . . and the nest be funden in the Trees at Beltane the trees shal be foirfaulted to the King . . .

Doves, on the other hand, were given special protection, penalties being imposed on anyone who destroyed dove cotes.

Linlithgow Palace would, at the hands of James's descendants, become one of the finest of Renaissance palaces, a gorgeous construction, galleried and fountained, embellished and carved and barrel-vaulted, the favourite of queens, birthplace of Mary Queen of Scots.

Whilst his son James II took no interest in the rebuilding of the palace, he appears to have inherited the unfortunate tendency towards prejudice against rooks, because on ascending the throne after the murder of his father he enacted further anti-bird legislation in 1457, widening the scope of the law to include other species:

> Pertrickes, Plovares, and sik like foules . . . eines, bissettes, gleddes, mittalles, the quhik destroys, baith cornes and wild

foulis, sik as pertrickes, plovers, and others . . . And as to ruikes
and craws biggan in orchares . . . and the nest be founden in the
trees at Beltane the tree shal be faulted to the King.

(Beltane is the right time, if one is intent on such purposes, one of the
old Celtic divisions of the year, more than just seasons, they are intervals of meaning between light and darkness, warmth and cold. Falling
on 1 May, Beltane or Bealtuinn is the optimum breeding time, not
only for rooks but for most other birds.)

Last spring, while I was buying an old book about crows in a shop
in Deeside, the lady selling me the book looked at the cover: 'They're
meant to be intelligent birds,' she said.

I agreed.

'Pity they're regarded as vermin.'

The word reminded me of the history, took me away from my own
small view of corvids – of corvids in ones, individuals – far in concept
from the idea of 'vermin' with its manifold suggestions of low-life
commonality, disease-carrying or wilful harm, its overtone of disgust,
its hints of justifiable destruction.

The killing of corvids, and indeed most other bird and mammal
species regarded as being in one way or another detrimental to human
interests, has been permitted by law in Britain for centuries, and for
longer in Scotland, to its discredit, than in England.

'Vermin' is a word that still sanctions all, explains and allows every
inventive, malign, brutal method of destruction, every way in which
birds and animals were and are trapped and netted and shot, ways

elaborated and refined through time (as humans, this seems to be one of our greater areas of expertise) and illuminating only the boundlessness of our own savagery, our feral cunning, our knowing less about our prey than about the methods of its destruction. With crossbows, arrows, in traps and snares of every elaborate and fiendish sort, with bird lime, a sticky substance smeared on branches to trap the unwary, with poisons – arsenic, mercury, strychnine – with modern pesticides – alphachloralase, cymag, aldicarb, carbofuran – by gassing, shooting, methods often involving something more purposefully cruel than mere disregard for pain, with our own knowledge growing in sophistication, we have damaged, often irrevocably, our native species.

In North America, it was, John Marzluff and Tony Angell suggest in their book *In the Company of Crows and Ravens*, the arrival of Europeans, bringing with them their old hatreds and superstitions, that introduced the idea of the undesirability of corvids, an idea that was to override and destroy (among the many other things in that territory that were erased, extirpated for ever) the protection extended by the traditional respect and mutuality of the indigenous inhabitants of the continent towards the birds and animals with which they shared the land, cultures in which corvids were, and still are, treated with greater degrees of respect than in Europe. Native American culture venerates ravens, admires crows. Magpies are honoured in Lakota dances because the black and white of their feathers represents both living and dead.

By the twentieth century, with true American enthusiasm for the task in hand, corvid colonies were being destroyed by dynamite. That this literal overkill, the use of bombing against birds, doesn't appear to have made any difference to corvid numbers can only be a comfort to those who might question, in general, the results of disproportionate balances of power.

It's easy now to judge and to wonder, at a point so distant in time, but more difficult to enter into the small area of the mind, territory, vision and hopes of many of those who lived in times both brutal and precarious. It may be unreasonable to expect people to treat animals with greater benevolence than they do other humans, especially in the face of the unbridled terrors of life in past centuries, plague and war and famine, with no resources to ameliorate the capriciousness of nature, weather, bird or beast. Nor it is difficult to summon an idea of anger or despair as newly planted crops, the sole guarantee of life over death, were raided annually, serially, by birds, or when the birds appeared to be gaining ascendancy in a perpetual war over naturally occurring food.

England was later in its embarkation upon the processes of permitting, indeed encouraging, the slaughter of birds and mammals than Scotland, passing the Vermin Acts a century later, in 1532 and 1566. The first applied to rooks, choughs and crows and, like the Scottish acts, was designed to stop their plundering of cornfields. The 1566 act was extended, inexplicably, to cover virtually every creature, bird or mammal, from mice to buzzards, that occurred naturally in England, and was given an added guarantee of efficacy by the inclusion of bounty payments on presentation of the corpse of the offending creature.

At least part of the unfortunate reputation of corvids has been based, historically, on incorrect observation, lazy assumption and, on occasion, pecuniary advantage. Corvids rarely kill healthy lambs. They do attack weak, dying or dead lambs. They will peck out eyes and tongues, eat placentas. Reporting of corvid damage by farmers appears to be notoriously inaccurate; many, observing crows or ravens in the vicinity of dead lambs, make the assumption that it is they who are responsible for killing, and whilst seeing corvids pecking lambs' eyes is not a sight most people relish, it isn't the same as their killing of lambs. In *Mind of the Raven* Bernd Heinrich writes of accusations made against ravens for killing cows and calves, the result of one report being the implementation in Arizona of a raven-eradication programme, which included compensation payments for farmers' losses. In Germany a few years ago, hysterical reports of 'gangs' of ravens killing calves appeared in the media. That the assertions were untrue, Heinrich suggests, is proved by the fact that ravens' beaks are unable to penetrate the skins of even quite small animals. (He provides a marvellous example of raven ingenuity. Some ravens, although unable to pierce the skin of road-killed squirrels, are able to scoop out the innards through the mouth, to leave only an empty, perfectly intact, inside-out squirrel skin.) If ravens, with their large, powerful beaks, can't kill calves, crows with their smaller ones can't either. Ravens will also deliberately, carefully test to ensure that a creature is dead by carrying out 'jumping jack' behaviour near it before approaching closer. In Germany, subsequent investigations (including post-mortems) demonstrated that most calves were either moribund or already dead when attacked. Once proof of raven-killing was required, government compensation was no longer paid.

Now, though, we know enough to be able to judge the balance between the harm corvids cause and the benefits they bestow. The crows and rooks who feed on young crops significantly reduce numbers of crop pests, lessening the need for pesticides.

The sight of crows feasting on the aftermath of the Great Fire of London was deemed so shocking – despite the fact that they were carrying out a task of clearing and cleansing that would inevitably have to have been done by someone – that the ensuing opprobrium damaged corvids' prospects for centuries, in the sorry equation that made their feeding on the dead almost worse than the deaths themselves. Battlefields, those bloody drifts of the wantonly killed, of men destroyed by their own, drew (and no doubt draw) crows to feed, yet human disgust is for the feeding birds, not for the pointlessness of war, for the instigators, the perpetrators, the paymasters, the makers of arms, or indeed for the apparently insatiable human desire for the often illusory attainment gained only by conflict. Corvids seem to absorb and reflect our guilt. We are casual in our waste of the lives of other humans but reverential in the treatment of their remains. The vital role of the carrion eater to the eco-system continues to be misunderstood, or disregarded. A ghillie in the West Highlands described ravens to a friend of mine as 'the housekeepers of the hill'.

The things we do to animals that are similarly, indeed more, unpleasant are done carefully out of public view, in slaughterhouses, and factory farms, the egregious places where intensive rearing and killing in all its cynical, venal brutality is carried out. (About the things we do to fellow humans, it would seem we feel not even that degree of shame or remorse, for those, as television news bulletins show, we do not even bother to conceal.)

In Scotland, it was the aftermath of the Jacobite wars, following the defeat at Culloden, that saw the most determined campaign against the natural world begin, when the disintegration and destruction of clan society initiated a relentless process of change and obliteration beginning with people and ending, inevitably, with birds and animals. The land, emptied of people, was overtaken by the ruinous advent of those apparently harmless, landscape-destroying, all-nibbling sheep. As the sheep destroyed the land, birds and animals deemed inimical to the interest of landowners and farmers – eagles, sea-eagles, corvids, foxes, polecats, martens – were themselves relentlessly destroyed. With the ties to land and obligations to tenants loosened or removed altogether, large-scale alteration in land use heralded the development of wild areas of Scotland as places where grouse and pheasant could be reared for shooting. As it was with sheep, so with grouse and pheasant: every species that threatened 'game' birds had to be removed.

Victorian fashion too brought wealthy people north to the newly expanded Highland hunting estates. During the Edwardian years, the gamekeeper came into his own. Their numbers were the highest they

eye-pecking, flesh-eating

would ever be, their duty the ferocious protection of the landowner's stock by relentless destruction of birds and animals that threatened it. Magpies almost disappeared from large areas of Scotland. Even now, Scottish newspapers regularly report the finding of poisoned birds – ravens, red kites and even golden eagles – killed in order to protect pheasant and grouse.

In 1746, Linlithgow Palace was destroyed by fire for a second time. The symbol of Stuart power was reduced in a night, to ash. On its lovely site overlooking the loch it remains, still beautiful, standing in emptiness and ruin, its walls and rooms open to the sky. It is, though, inhabited again. By a strange, suitable turn of corvid fortune, jackdaws live there instead of Stuart kings. They fly through the ruins of the Lion Chamber. They dive with their black and sinuous flight, swoop, loop, spiral through lancet windows and oratories, round mural gallery and clerestory, through the empty rooms of James's vision. Time changes all things, if not always for the better, then sometimes at least; Linlithgow makes me hope there may be a day when some humans will no longer think it desirable, or even acceptable, to kill one species in order to protect another, specially nurtured, carefully bred for the sole purpose of being shot for sport.

The fear that maintains the invisible barrier between human and corvid (on the corvid's side, at least, entirely justifiably) is more than just physical fear. For humans, large, noisy, powerful though we may be, the feeling is real and complex, fear both of what birds might do and what they may or may not represent. Beyond the physical is a darker fear, more profound and atavistic. It is a sense that may share roots with arachnophobia, living in us deeply, perhaps so ancient in its origin as to stem from our and their dinosaurian roots. It persists in words and ideas, in culture and memory, in folk tale and nursery rhymes, in the famous Scottish ballad 'The Twa Corbies', with its eye-pecking and flesh-eating, its final, memorable image of an eternal wind blowing through picked and empty bones; in the Grimms' version of *Cinderella*, the ugly sisters having their eyes removed (perversely enough by pigeons); in swarms of pursuing malevolents in *The Birds*, absurd though the vision of the birds in Hitchcock's film actually is, as is the story on which it's based, Daphne du Maurier's dark, apocalyptic tale, seen by some as a prescient herald of coming environmental disaster, unfortunately using, as the symbol of foreboding, species more victims than instigators in the irreversible processes of destruction and decline.

Superstition blends with cultural association, undiminished until now. In the north-east of Scotland, corvids are associated with people who practise 'the black airts'. (I don't know what the black airts involve but they sound darkly enticing.) Corvids are reputed to form pacts with the Devil. I look at Chicken. I can't imagine what she might gain from the association, but if she does, she keeps it very quiet. In an e-mail, a friend tells me of the sighting by a mutual friend in the north-west of a raven in his garden. The sighting was deemed, a posteriori, to have been one of ill-omen because of the discovery a few days later of the illness of a near relative. I reply in robust and forceful terms. I suggest that the raven sues. (I'm greatly cheered by discovering that rooks, according to one legend, escort the souls of the righteous to heaven, a pleasing thought indeed for the righteous, rooks being kindly, sociable types.)

The story that, for me, illustrates most powerfully the force of superstition was written by Truman Capote in 1964, about his relationship to his raven, Lola. For all the beauty of his writing, the characteristic humour and sharpness, I find difficult it to read. The account, which appears in *A Capote Reader*, was written twelve years after the winter when, living in Sicily in the early 1950s, Capote exchanged Christmas presents with the village girl, Graziella, who came to his house every day to clean and cook for him. He gave her a scarf, a sweater and a necklace. She gave him a fledgling raven that she had caught, with considerable effort, in a ravine in the hills above Bronte in Catania. Capote, who writes of his previous dislike and fear of birds, describes the raven, as he first saw it, as 'both dreadful and pathetic', with severely clipped wings, 'black beak agape like the jaws of an idiot, its eyes flat and bleak'. With revulsion, he shut it away in a spare room, visiting it with reluctance. Only in spring, when one day the bird disappeared, did he realise, through his sense of loss, his affection for her, deciding in an instant that she was Lola, a name that 'emerged like a new moon overhead'. In panic, thinking of her possible fate, he and Graziella searched until Lola was found in an unused room, the moment from which she became integrated into Capote's household, ruling his two dogs, taking advantage of them, stealing food, riding around the garden on the bulldog's back. The bird thought, Capote writes, that she was a dog: 'Graziella agreed with me, and we both laughed; we considered it a delightful quirk, neither of us foreseeing that Lola's misconception was certain to end in tragedy: the doom that awaits all of us who reject our own natures and insist on being something else than ourselves.'

Lola was, like all corvids, inclined to steal and cache. Among her prizes were the false teeth of an elderly guest, who was very upset, obliging Capote to find where it was that Lola cached her treasure:

She leaped from floor to chair to bookshelf; then as though it were a cleft in a mountain leading to Ali Baba's cavern, she squeezed between two books and disappeared behind them: evaporated like Alice through the looking glass. *The Complete Jane Austen* concealed her cache, which, when we found it, consisted, in addition to the purloined dentures, of the long-lost keys to my car . . . a mass of paper money – thousands of lire torn into tiny scraps, as though intended for some future nest, old letters, my best cuff links, rubber bands, yards of string, the first page of a short story I'd stopped writing because I couldn't find the first page, an American penny, a dry rose, a crystal button . . .

Two tragedies were to change the lives of both Capote and Lola. The first was the stroke suffered by Graziella's father, the second, occurring the following day, the accident in which her (somewhat disreputable) fiancé, Luchino, knocked over and killed a child, ending his and Graziella's prospects of marriage. Graziella blamed Lola, saying that she was a witch and had the *malocchio*, the evil eye, that what had happened had been brought about by the bird as punishment for her capture. The taint of *malocchio* spread to Capote, with Graziella and everybody else refusing to enter his house. People in the streets, on seeing him, crossed themselves and made the sign of the horn. After having lived there for two years, Capote decided in a single night to leave. He packed the car with belongings, dogs and bird and drove on a journey marvellously evoked, with Lola on his shoulder, to Rome.

The story, like all Capote's writing, has profound qualities, light undershot by darkness, of dazzle foreshadowed, the dangerous edge familiar from *Breakfast at Tiffany's*, from Holly Golightly, a slightly less charmed ingénue in the book than in the film (who, in her former incarnation as Lulumae, fourteen-year-old bride of the luckless, elderly Doc Golightly, was given as a love token the crow he had tamed and taught to say her name). What it doesn't have is the hard-edged glint, the knowing bitterness that underpins so much of the work as it did the life, the prevailing sense that trust is foolish, love evanescent.

The language Capote uses of Lola is the language of admiration, bewildered affection. Writing of their subsequent life together in a fourth-floor apartment in Rome, he describes Lola's pleasure in sun-bathing on the balcony, of providing her preferred mineral water for bathing, of the admiration too of his ninety-three-year-old neighbour Signor Fioli, who liked to watch Lola, smiling if she did anything 'foolish or lovely'. It is as if, in this unlikely fellow inhabitant of the earth, a sad and complex man formed one of the few bridges that spanned the overwhelming loneliness of his life.

It is the ending of the account that is terrible. Frightened by a cat, instead of flying, Lola jumped from the balcony, landing on sacks on the back of a lorry, to be conveyed away. Capote describes his rushing down six flights of stairs after her, calling, falling, losing his glasses, which smashed against a wall – how, running, blinded and crying as the truck turned the corner, he didn't quite see as Lola disappeared from his life for ever; and I find it almost unreadable, thinking of the shadow of the *malocchio*, the blighted future for both man and bird.

Image, as we know, is all. One particular poster for a horror film shows a crow – a crow not doing anything very much, standing, looking as crows look, neither threatening nor ominous, looking merely crowish and utterly benign, but it needs do no more. Its portrayal is enough.

After people have visited, wondered, commented, shown their fear, I let them go on their way, perhaps slightly enlightened: the men who come to check the gas boiler, the coalman, the lady from one of the Scottish wildlife organisations, visiting on fund-raising business, who seemed in Chicken's presence as bemused, as uneasy as all the rest. I donated money. She left some leaflets which, shortly afterwards, Chicken removed from the chair in the study where I had left them and shredded before stuffing them into the hole in the kitchen wall where she keeps her important documents.

It may, too, be the notion of wildness that is frightening.

'Is she a wild bird?' people ask. I don't know what to say. When I think of the concepts of 'wild' and 'tame', they seem wrong together, the words ill-fitting. They are not opposites, not antitheses. I try to fit the terms with what I know of the birds but can't. They may be 'wild' depending on the word's own definition of itself but they are neither tame nor tamed:

> **tame** lame, cowed, de-natured, docile, neutered at the core. Subjugated.
> **wild** living in a state of nature, inhabiting wild places, not tamed or diminished, savage, ferocious, not submitting to restraint.

I realise that if 'wild' was once the word for Chicken, it still is, for nothing in her or about her contains any of the suggestions hinted at by the word 'tame'. Chicken, Spike, Max, all the birds I have known over the years, live or lived their lives as they did by necessity or otherwise, but were and are not 'tame'. They are afraid of the things they always were, of which their fellow corvids are, judiciously, sensibly: of some people, of hands and perceived danger, of cats and hawks, of things they do not know and things of which I too am afraid.

'Not tamed or diminished'. I walk past two crows paying attention to some dropped food on a pavement. They notice me but carry on. I pass them at an appropriate distance. We co-exist and do so because we have both learnt necessary boundaries, theirs the boundaries of fear. Where does wildness begin and how far does it extend? It's more than what they are, what we are. Wildness is a continuum. Swifts or terns or albatrosses are wild because there is no point of meeting between them and us, but for other birds, the ones who live in greater proximity to man, their wildness is other, knowing, watchful.

In *Arctic Dreams* Barry Lopez describes walking one evening on the Ilingnorak Ridge, in the Brooks Range in Alaska, among tundra birds, the calm stares of the horned lark, the Lapland longspurs, the golden plovers which abandoned their nests at his approach; of the snowy owls, the one that looked into his eyes when he began to move. He writes of their 'wild, dedicated lives', birds that have little or no knowledge of man, living in a place where man's influence may be insidious, even pernicious, but is not immediately obvious to the creatures who live there. I find it moving, the extent of their wildness, the lark who stares back 'resolute as iron'.

Domestication is a different process, a long one, the true process of 'taming' for humanity's benefit or use, but even birds reared in captivity – the parrots and parakeets – seem to relinquish as little as they can of the essence of themselves, far from their long and terrible experience of trapping and trade, unaffected by the destruction of their fellow species and, now, the destruction of their natural habitats. In this at least, corvids are fortunate. Unlike parrots, whose brilliant, vibrant beauty and astonishing vocal abilities have made them vulnerable, corvids' beauty is subtle, their voices lacking the qualities to make them prey to human traffickers. All the birds I've encountered, amenable or not, intelligent or not, capable, all of them, of far more than we have yet discovered, have been, always, their own birds.

Barry Lopez's words resonate, make me realise again what I already know, that they are all like this, none of them tame, all of them resolute as iron.

Often, I measure the distance we have come, Chicken's acceptance of people, my own, still sometimes surprising, ease in the presence of birds. I look at the crow on the horror-film poster and think it sweet. Now, I see others' fear and I almost forget my own.

7

Spike

There are traces of him still around the house, perceptible, often indelible. There is, on the fly-leaf and the first pages of my hardback copy of Anne Michael's *Fugitive Pieces*, a faint, penetrating stain of red-brown, a bloodstain, where a beakful of mince was hidden. There is the large paper butterfly I have kept, a Red Admiral, concealed on top of a kitchen cupboard, discovered only recently when the room was being repainted. There is the feather I have suspended on a thread above my desk lamp, the feather that turns slowly, glowing marvellously, enamelled blue-gold in the warmth and light. Signs of various and varied depredations remain, uneven pieces torn from the lower edges of wallpaper near the skirting board in the spare bedroom, the occasional stain on a carpet, many forgotten rips in book pages which I remember only when I take the book from the shelf.

On most days I drive or walk past the tree where Spike was hatched, the tree from which, gracelessly, he fell or was pushed by parents who perhaps saw in him a deficit I never did. In a street at right-angles to my own, the tree, a towering, bare-stemmed monkey puzzle, *Araucaria araucana*, stands in the garden of a house then owned by friends, a tree that must have been planted, with Victorian pride, sometime in the 1880s when the house was built. The houses in that street are like all the houses in the district but larger, detached, grander, the gardens stretching on the south side to lanes behind, lined with the remnants of the coach-houses where once carriages and horses were kept, now garages, and on the north side down through fringes of trees to the waters of a small burn that flows fitfully through the city.

There is another old monkey-puzzle tree in the area, even nearer my house than my friends' tree, old enough for the stiff, dark branches to dip low, sweeping towards the ground, in the front garden of a Queen's Road mansion, now turned into a social club. For all the years I've lived here, I've watched the progress of magpies to and from this tree, two, three, four of them at a time, darting with shifty vigour into the dense centre of stiff, symmetrical branches, towards a nest barely visible in the dark centre of the tree, or emerging, slightly startled, into the cold light of Aberdeen day. The nest is almost completely concealed. Only the glimpse of an untidy trail of twigs hints at an elaborate construction behind. Magpies' nests are large, layered, well designed, executed with care against weather and predation. They can take longer to construct than many modern houses: weeks spent transporting twigs, mud, grass, forming them into deep, domed super-structures, lining the curved sides with feathers, sticks, hair, *objets trouvés.* Some are dome-roofed, accessible by side entrances, magpie cathedrals, magpie palaces; all, I like to imagine, fan vaulting, Romanesque arch and *piano nobile.* In the Hunterian Museum in Glasgow, there is a magpie's nest constructed, along with the usual assemblage of twigs and leaves, from metallic objects, old coat hangers, builders' waste, barbed wire, a glittering object of strange charm and beauty.

From my bedroom window I watch magpies, flying, criss-crossing, I believe, from one old tree to the other, over the steep Parisian-style mansard roof of the house opposite, as members of two magpie families (clans, perhaps, Macdonalds and Campbells, Montagues and Capulets) or one extended family exchange visits, for what purpose, peaceable or otherwise, I do not know.

For a long time I liked the idea of magpies, admired their beauty and their flight, their presence always seeming to me new, unusual – perhaps because, outside their depiction in books, there were no magpies in my childhood. Had there been, I'd have remembered. Their jewelled brilliance would have lit the grey Glasgow of my youth, in gardens, during walks in Pollok Estate where they would have added a gilded edge of magic to an already-enchanted child's world of clear streams, small wooden bridges, conker trees, or in Maxwell Park, flitting over the enclosed hollows of privet hedges – but there were none then, in those days before they began the steady move north, at least before they reached the Scottish cities.

In a book on the wildlife of this city I read that, for a long time, there were few magpies in Aberdeen. They began, it is suggested, inhabiting suburban locations in the late 1940s, moving over the decades to establish themselves in urban parks and gardens. Where were they before that?

I ask a friend, a native speaker, possessor of a fine Doric voice, if he can tell me the Doric names for crows and magpies. Doric, the language of the area, of north-east Scotland, is the language of Aberdeenshire, of Buchan, Fraserburgh and Peterhead, of the country to the south of Aberdeen, round the Mearns, farming country, deep-red-earth country, rich, green tree and field and hill country, the country of Lewis Grassic Gibbon's wonderful trilogy *A Scots Quair*. It's mainly a rural language, named, possibly ironically, for the Greek rustic tradition, a counterpoint to the sophisticated Attic traditions of Athens, translated for Scottish use into a comparison with snooty, sophisticated Edinburgh, 'the Athens of the north'.

There are, my friend says, *cras* and *corbies*, *pyots* and *pies*. Magpie, and indeed corvid, history must be long in the area, for were Doric to innovate linguistically, it would not be in order to invent a name for magpies. (In Doric, a common greeting was, and still is, 'Foo's yer doos?', F being commonly subsituted for Wh or H. The question suggests that if your doves are well, it may reasonably be assumed that you are too. The required response is 'Aye peckin.')

Not only Doric, Gaelic too has names for them all: *cathag* the jackdaw, *pioghaid* the magpie, *rocais* the rook, *feanna* the crow, *fitheach* the raven, the raven that, before this spring, I'd only seen once, high on Na Gruagaichean in the Mamores, gliding serenely, darkly on a bright summer afternoon with his three companions into the blue-white mist over the mountain. Gaidhealtachd, the area where Gaelic is spoken, used to extend south and east, far beyond its current territory of the north-west Highlands and the Western Isles. Language, the custodian and mirror of history, seals the past in its words, tells of what there was, if not what there is; the Doric word for magpie, and the Gaelic too, telling that once magpies were abundant in northern Scotland. Although the names were there, for many decades it was only the memory of a presence that remained because, in the main, *pie, pyot, pioghaid* had gone.

Magpies were common even in the north of Scotland before the relentless persecution by nineteenth-century gamekeepers reduced their numbers severely. Over the decades since the Second World War, their numbers have increased and they have migrated south to north, country to town, spread, become, in their uncommon way, common, although to me they still seem exotic, almost out of place, wilfully glamorous, super-confident among the quieter-toned brown and black birds of our environs, traversing houses, roofs, chimneys with their joyous, undulating flight, soaring, dipping triangles of frill-edged wings and iridescent tails.

Knowing that there were magpies in my friends' tree, I asked, on a visit to their house one winter, if they would let me know if they ever found a fledgling that had fallen out of its nest. They said they would, but the way they said it made it clear that they considered the request one that might be made only by the frankly deranged. I asked, I realise now, without thinking that it would happen, without considering what I might do if it did. At the time of my asking, we still had plenty of other birds in or around the house. They were by then mostly long-time residents, some, in bird terms anyway, already elderly, Max the starling, many of the doves. Chicken and Bardie, at around five and eight respectively, were still in human terms pre-pubescent. About the age of Marley, recently acquired, we knew nothing.

They've left now, my friends, Elizabeth and William, the ones who owned the house, the tree, the nest of magpies, going to live in Edinburgh when they retired from university, from distinguished academic careers as teachers of divinity. A suitable household from which to acquire a magpie.

It was in spring of that year, a morning in early May, when Elizabeth phoned. A magpie had fallen from the nest. If I wanted it, I had better be quick because the neighbour's cat was taking a close interest. I snatched a brown-paper bag and drove the two hundred or so yards in my anxiety and impatience to be first to reach the bird. It was still there, an object on the sparse grass under the tree, barely recognisable as a bird. It neither walked nor flew. Rather it rolled, fast, a movement more like that of the joke wind-up spider I have in my desk drawer – a roundel of black fuzz with a key, two protruding legs and glued-on eyes which, when wound up, rolls maniacally from under the bedclothes, or wherever it's been concealed – than a bird. Whatever it was, I chased and it rolled, for several embarrassing minutes, round and round the trunk of the tree, watched from the top window by William, an undignified progression which felt, and was, unequal until, eventually getting up the speed or wile necessary to gain on a three-inch fledgling, I grabbed the thing and put it into the brown-paper bag. Together, we drove home.

Transferred from paper bag to table, the thing was definitely a bird. Barely feathered, the faintly contrasting fuzz on his body the sole indication that he was a magpie. He was as all fledgling corvids, a pimpled, boggle-eyed *joli-laid* in a wispy suit of down and pin-feather, a covering of faint black and white overlying the pink; yellow-gaped, the delicate primrose crown of infant feathers standing up like a mad crown around his head. He looked at me, this infant – as Chicken had been, only a couple of weeks from the egg – with calm and perfect equanimity, assessing me as completely, as thoroughly, as I was assessing him. A magpie. I put him into a newspaper-lined cardboard box, deep enough to confine him, not sufficiently deep to prevent him from seeing what was around him.

Magpie. *Pica pica.* The name alone conjures demons, prises open narrow apertures into deep wells of superstition. Even to reflect momentarily on the word 'magpie' is to burden one's mind with other people's fears, to unravel a Western ragbag assemblage of all that's non-scientific, primitive, atavistic. I assembled some of the names I knew, the stories, the omens, the auguries, thought of the careful, time-worn formulae that might help prevent or ameliorate the terrifying damage this tiny fledgling could perpetrate, just by being; the words, greetings, kind and degree of respect that must be shown to those of mystic power in order to appease, to soothe or flatter or otherwise take the creatures' minds from the evil, myth and legend suggest they carry with them. 'Good-morning Mr Magpie!' I would have to cry, presumably frequently if we were to share a house. I would have to cross myself, or spit over my right shoulder and say (or better shout), 'Devil, I defy thee!' I would, according to one suggestion, be wise to carry with me, at all times, an onion. Names from books of folklore stacked up: chatternag, chatterpie, haggister. The counting rhymes: 'One for sorrow'. Ordinary words of description were no better: 'noisy', 'hectoring', 'thieving', 'aggressive', epithet and accusation, this long history of guilt. 'Scandalmongers', Ovid called them, with what justification or provocation I don't know. A creature, according to Scottish legend, that carries a drop of the devil's blood under its tongue. (An unsurprising idea for Scotland, a place that regards eloquence with equal fear and mistrust.) What was it about a magpie that brought on its very small, pristine head this sustained vocabulary of opprobrium? 'Bird of ill omen'. I looked at him. He opened his pink, gilled beak, squeaked, flapped his unfeathered wings, already hungry.

I took out my falling-to-bits copy of Kenton C. Lint, to refresh my memory, resisting the desire – at least until I had fed the bird – to stop and read, to find out what lammergeyers should eat, wreathed hornbills, thick-knee Australian tanagers. I fetched out my ready and waiting box of all-purpose bird food made specially for insect eaters, a kind of bird muesli of strangely enticing smell, reminiscent of juniper, sumac, a bit like Marmite but nicer, a substance you'd like to taste but don't quite dare to. I mixed the stuff up with water into a gritty paste. The bird, a considerable gourmand from the beginning, ate with gratifying enthusiasm (as well he might: the ingredient list rivals in its variety anything sold by Fauchon, by Dean & DeLuca; fruits, seeds, aerial insects, honey, oils, shrimps, vitamins – more vitamins than I had known existed). After this, his first human-administered meal, he settled into the corner of his box and dozed. I went into the garden to cut a small branch to make into a perch.

'Thieving'. 'Aggressive'. 'Cursed'. What was it about this bird? Could it be as simple as the fact that magpies are black and white? It seemed too much for one small bird to bear, all that's contained, all that's implied in the cultural and religious canons of Western civilisation, the symbolic, iconographic poles of culture and ideas: black and white. Heaven and earth, life, death, good, evil, light, darkness, all things fundamental, elemental, *reductio ad absurdum*, a universe of fears combined to obscure the evolutionary process that delivered this startling bird into a world that appears still unready for it. Even mythic explanations of magpies' attire do them no favours; magpies, being black and white, are cursed by God because they did not don full mourning after the death of Christ. (An alternative version is that the magpie refused to console Christ, as the other birds did, during the crucifixion.) The magpie alone refused to go quietly into the Ark, instead standing on its roof shouting and swearing for the duration of the Flood. So, even by the time of Noah, the magpie had already made some serious enemies, the weight of the Church against him although his dress is monastic, black hood over white, a tiny, hobbling

a tiny hobbling Dominican

Dominican. In an unsurpassed, unsurpassable slur, French legend suggests that evil priests become crows, evil nuns magpies. Is the magpie's black hood that of the Reaper himself? Not that a magpie's black feathers are black. They're electric blue, violet, emerald, they're overlaid by gold, by bronze, the dazzle of shot silk and taffeta. The white of his chest dazzles. It reflects almost blue, like light on silver or on snow.

Magpies seem, oddly enough, to have been involved, certainly more than the average person or indeed bird, in discourse with the Almighty. A twelfth-century bestiary, *A Book of Beasts* translated by T. H. White, reports a magpie's conversation with God: 'I Magpie, a talker, greet thee Lord, with definite speech, and if you don't see me, you refuse to believe that I am a bird,' and indeed God seems to have taken a personal hand in disciplining magpies over a variety of indiscretions, from ignoring dress code to interfering in others' affairs by usurping the power of prediction. Might the magpie be something other than he seems? Satanic magpie? Heretic magpie? Cathar, Albegensian, Manichean, a representative of duality, this small, bright bird, hovering between God and Satan. (Thinking of it, it seemed strangely appropriate that he should have come from the background he did.)

China, paradoxically – a nation not universally acknowledged as a haven of animal rights and protection – is kinder, in literature and myth at least, to magpies, and to the other birds in classical Chinese poetry that represent all that's wistful, longing; images of distance and of love are summoned in the subtle, exquisite allusiveness of T'ang poetry, and when Li Bo writes of 'autumn wild geese', of 'cold crow perching on a branch', his readers know that he is writing of the sorrow of departure and the pain of separated lovers.

Part of the character for 'bird' – radical 196 in that most elaborate of linguistic orderings by which it's (almost) possible to systematise all Chinese characters and thus look them up in the dictionary (by way of identifying the radical and counting the number of pen strokes that make up each character) – appears in the names of most birds, including magpies. The name for a magpie is *xǐ qùe*, 'auspicious magpie', a welcome, graceful reverse of all the malicious, nasty, gossiping maledictions of Western anti-magpie ideology and propaganda. The Chinese equivalent of Valentine's Day, in the seventh month of the lunar calendar, celebrates two parted lovers reunited for a night by a bridge of magpies in the sky.

Black and white do not hold the notion of opposing duality in China that they do in the West. Yin and yang, the two principles that order the earth, are represented as black and white, unified in a symbol of wholeness – yin, all cold, shade, moon, female, is only the obverse, the partner, of yang, sun, heat, brightness, male.

That spring, Han was still at home, in her last year at school. When she arrived back, we took the magpie out of his box. His eyes, as Chicken's had been, were not yet black. His were grey, fringed by dark, feathered eyelashes. 'We can't keep him, you know,' she said, bowing to the good sense that told us both that we had enough birds to look after, that if there was an alternative to keeping him it should be sought. We both knew that we would not find it easy to part with this enchanting bird. After our conversation, I phoned Kevin the bird man who, years before, had so sagely and with such success chosen a suitable mate for my bereaved and grieving dove. We discussed the matter of the magpie.

'Lively birds, magpies. Need a good thirty feet to fly in,' he said. 'Bring him to me when he's ready.' We didn't discuss whether Kevin would keep him or release him. I measured the length of the kitchen and rat room.

How was I to know when he was ready? Did I ever intend to let him go? I can hardly remember what I felt then. I knew the facts, the same doleful tally with which, over the years, I had comforted myself about the nature of Chicken's existence, the ones concerning longevity in the wild, the hazards of reintroduction to the wild, the facts seeming to point only to the evanescence of the lives of birds. Uncertain as to the future, we named him none the less, a name of Han's choosing. Spike. The name was good. It felt appropriately pugnacious, containing everything one might have wanted to convey, a certain raffish creativity, an energy, with the echo of his Linnaean designation, *Pica pica*.

Han and I bathed Spike, or rather allowed him to bathe himself. We gave him a wide, shallow dish of water which we put in the bottom of the deep Belfast sink in the rat room.

Bird infancy is short. Spike snoozed, woke, fed, like an infant from a well-ordered nursery, and over the days of spring began to grow into his role and status as magpie. Fluffy, still-thin black and white feathers began to blossom from the pink of his skin. He began to show an interest in flight, launching himself from his box, from the edge of a chair onto which he had managed to jump, and on one occasion from an open window. (We were sitting in the garden below and managed to apprehend him as he plummeted, clearly not yet ready for flight.) He was amiable, bright, watching, with the confident air of the child who requires telling only once.

Early, we began to notice the ways in which Spike was different from Chicken who, as an infant, enjoyed a brief interlude of wayward behaviour, soon adopting the settled, staid habits she has retained ever since. Spike was vigorous, enquiring. He climbed, hopped and, eventually, flew. His ability to climb and leap allowed him to negotiate any obstacle with ease, floor to chair to top of fridge to kitchen cupboard, all without flying. He did eventually learn to fly the length of kitchen and rat room, a reasonable distance for an indoor magpie, slightly more than the designated thirty feet or so of clear flight-path. He ran, fast. He hopped. He ascended the stairs by hopping, descended by flying. He still rolled, when chased. I have no idea if this is common magpie behaviour or was unique to him.

Spike's eyes were different from Chicken's. Chicken has a slight, modestly pearly nictitating membrane, the protective cover that can be drawn over a bird's eyes at moments of need, obvious in her only when she's nearing sleep. She has too the hint of a feathered line across the eyelid, a feint giving her the appearance of being awake when she is in fact asleep. Spike's nictitating membrane was flagrant, a crazed curtain of saffron yellow which he used as much for expressive purposes as protective, rolling back his eyes, pulling down the yellow hood to demonstrate frustration or contempt, as a teenager will raise their eyes heavenward in inchoate rage or despond at a parent's failure to grasp the essentials of the given argument. Delight too caused him to shutter his eyes, the prospect of mince, the sight of a tub of chicken livers, a morsel of raw fish, all would induce an eye-rolling, yellow moment of anticipatory relish.

I don't remember when it was that I began to appreciate the capacities of magpies, when I began to realise that the prejudice, superstition and myth are based on observation, but observation misinterpreted, on beliefs so misguided that, in their judgements, generations of humans have mistaken acumen for evil.

I noticed in every natural-history book concerning corvids photographs of magpies fighting, either fighting one another or squaring up to something of disproportionately larger size, usually terrifying: a buffalo, a bison, a wolf, a fox. In one photo in a book I have, three magpies surround an irritated-looking golden eagle, interrupted at his bloody dinner of an unrecognisable object lying in the snow. They have a *Well, what are you going to do about it?* air of chippy thuggishness; in another, a group of magpies forages casually among reddened ribs and empty pelt, watched by a patient coyote, his muzzle and chest stained rufous by the blood of his disappearing meal. Every representation expresses remarkable confidence, an *Are you looking at me?*

blend of smallness and pugnacity. (Spike, though, was frightened of some things: ladders, sparrow-hawks glimpsed through the window, flying dot-sized in the sky, a long striped red and yellow fuzzy object on a stick – a child's toy we called the Portsoy snake, after the small north-east harbour town at whose annual boat festival we bought it.)

But if magpies are aggressive (and I write as one whose sole, incomplete piercing – that of an ear – was carried out by a magpie) so are doves, and the word is not usually applied to them. Butterflies? Visiting the butterfly house in Amsterdam's botanical gardens, I watched, amazed, as one butterfly raised its wing, much as I've seen birds do, both to me (usually when I'm in the dove-house removing eggs in an attempt at post-coital contraception) and to each other, and struck its neighbour. Magpie aggression is only in the nature of our own aggression, territorial, sporadic, to do with the essentials of life: space, sex, food. It has nothing to do with intoxication, greed, revenge and the other dismal range of human banes. There are days when, contemplating the news, it seems worse than simple irony that we should dare to call magpies aggressive.

'Thieving'. *La Gazza Ladra*, 'The Thieving Magpie'. The word in itself suggests intent, a quite inappropriate application of human morality, consciousness of culpability, definitions of 'yours' and 'mine'. I never believed it was from the intent to steal that Spike would, if I injudiciously left my handbag in an accessible place, nick the odd tenner. Simply, he liked paper. The ensuing fight under the kitchen table indicated more venality on my part than on his. Professor Tim Birkhead, magpie expert, says that he's never come across shiny things in a magpie's nest, that it's observations from pet birds that lead people to the belief that they are attracted to jewellery. Spike, while delighted by glitter, was by flowers too, children's plastic teacups, hair ornaments, ribbon, coins, letters, silver foil. Magpies, like all corvids, cache. A cache of mince, a cache of bread dough, of stolen prawns, a cache of torn-up pieces of the pages of a book of poetry, a cache of oddments that might come in useful later. A cache of all things considered, with an avian eye, beautiful.

He passed from infancy to the equivalent of a toddler, a clearly intelligent, nosy, meddling, small, flying toddler. He outgrew his first cardboard box and we assigned him a larger one through which we cut holes and slotted a branch as perch. We put him in every evening. When he learned to fly, he'd fly there himself. 'Spike, bed!' we'd say and he'd fly the length of the kitchen, dive into his box and settle down. We'd fold over the flaps and close him in. Like a toddler, he'd escape sometimes and appear triumphantly in the kitchen, the avian equivalent of 'I want a drink!' wailing complaint as he was thrust back.

I no longer remember the point at which Han and I again discussed Spike's future, whether he was fully feathered or not, whether he was able to fly. We were standing by the kitchen stove discussing it, when we might think of taking him to Kevin's. Spike formed a triangle with us, part of the conversation, standing, as he had just learned to do, on the ears of the wooden rabbit on the mantelpiece, when he joined the discourse, gave forth his opinion, sealed the argument.

'Hello!' he said very suddenly, loudly, with astonishing clarity. Han and I stared, gaped. Then, even louder: 'Spike!' He was pleased with his effort. 'Spike, Spikey. Hello! Spikey? SPIKE!' His voice was a voice so human as to be shocking.

Our discussion ended. We had, whether by neglect or design, kept him too long. Like the child taught by Jesuits before the age of seven or one of Miss Jean Brodie's pupils, he was ours for life.

8

The Pleasure of the Bath

Always the most fastidious of birds, hygiene – or possibly cosmetic considerations – was of the greatest importance to Spike. Bearer of a set of sparkling white breast feathers, he ensured, with some effort and attention, that they were never less than perfect. On the rare occasions when his feathers were besmirched in any way, he became distressed; as when, during an attempt to drink the water from a vase of lilies (accomplished by clinging with his toes to the edge of the vase), the white of his feathers became stained bright pollen-yellow. Those who have encountered the persistent staining power of lily pollen will appreciate the near-miraculous removal, within ten minutes, of every sign of pollen from his feathers, a feat accomplished by a frenzied, earnest engagement of beak with feather, accompanied by low muttering and squeaking as he expressed disgust and displeasure.

Many birds appear to bathe for social reasons, or for enjoyment rather than for solely hygienic purposes. For Chicken, it may be either, as she bathes with the cleansing fervour of the *pujah*, of the Hassid at the Friday-afternoon *mikveh*, in a process of purification that recalls every notion of religious and cultural concepts of keeping at bay the mental and physical dangers of impurity: *haram, tehora, niddah, mahrime.* Both the timing and the thoroughness with which it's undertaken reduce one to the humbling suspicion that the purpose is not only one of cleansing, but there might be a symbolic purpose too, that of the removal of the soiling contamination of humanity.

Chicken's water dish is a large bowl with the word DOG insultingly integrated into the light brown of its side, the only vessel sufficiently heavy to prevent her from tipping it over and sufficiently wide to allow her to bathe with the ease she requires. With grace she steps in, bows, bends, lowers herself into the water, flaps her wings. Cascading drops fly across the room, drenching the wooden floor, towards the furniture, the books on the low table, me. Grunting and creaking with the delight of it all, she digs her beak deeply into the furthest reaches of wing and chest, through every feather, searching, scouring, before hopping out and shaking herself beak to tail (distributing a further volley of droplets to the extremities of the room), after which she repeats it all, once, twice, thrice. Stretching, she lifts one leg behind her, the wing on the same side extended backwards, a balletic posture of elegance and grace. Scratching is less so. She crouches, raises her leg to extend over her wing, her beak open as she scratches her face, the hard sounds of nail on tough skin, *crik, crik, crik, crik.*

There follows a post-bathing period of reflection and ease, a moment of luxury as she hops onto the sofa to stand on a cushion of her choosing, where she rests awhile, shaking herself occasionally, until she is dry. At other times she prefers, while still soaking, to jump onto my foot under my desk, grunting. She shakes herself vigorously, sifting her feathers one by one through the cleansing sieve of her beak, continuing to shake herself from time to time, showering the remainder of the water across my legs. A waft of wet feather seeps up from under my desk, the inimitable scent of damp rook, almost wet-dog but not quite. The smell is stronger, sweeter. The sound she makes as she cleans her wings is a long, rolling, satisfied half-grunt, half-growl, *wrrrrrrrrrrrr.* Being dry is more than a matter of comfort. The weight of clinging water hampers wet birds in their attempts to fly, making them vulnerable to predators, and whilst Chicken might be free from this particular threat, she none the less carries out the behaviour that has kept her species safe.

In most gardens where there is water in any receptacle – birdbath, puddle, wheelbarrow – small birds will be seen in the midst of a frenzied spray of drops. The doves will bathe outdoors on fearsome days of grim northern winter, days when sleet, snow and rain vie for precedence, when daylight wins marginally, briefly, over total darkness. I can't imagine why but these days are their favourite, their most chosen of days, when they rush out of their house to line up in unruly progression beside their washing-up-bowl bath, bickering for their turn to splash and bob in the ice-cold water. In their eagerness and excitement they jump on each other's heads, bite and peck, squeezing themselves into any empty corner; then afterwards, as if reclining on sun-loungers in a spa resort, they lie siren-style on their sides on the rain-slicked slates of the roof, often streaming with half-melting hail, raising their wings to allow further ice-cold water to penetrate to the very roots of their feathers.

On days when I'm indoors and wish to be nowhere else, in late January, February, when Chicken and I hide from the weather, when wind leaks upwards eerily through the floorboards, through the invisible gaps at the margins of windows and doors, they're lining up, capering in the queue under the steady fall of rain and semi-liquid ice, squabbling and hopping with perverse, determined joy.

Preening too is a major preoccupation for birds. All our birds – parrots, rook, magpie, doves, starling – have been dedicated preeners. For wild birds, keeping their feathers free from parasites is vital and though Chicken cleans herself in the same way, with the same frequency, she does not have, and never has as far as I can tell, the insidious population commonly inhabiting the feathers of rooks and other birds in the wild, the hippoboscids; the louse flies, flat and insinuating; the mallophaga, the chewing lice; the mites and nematodes; the gapeworms, a particularly nasty kind of worm that attaches itself to the trachea of birds, preventing them from breathing.

In addition to her own grooming efforts, Chicken requires a little help, the odd trimming of an over-long toenail, frequent attention to her beak, the upper part of which grows to protrude beyond her lower one, making it difficult for her to pick things up. When it's too long (although she anticipates it with scary prescience and runs away to hide) I attempt to take action. Clandestinely, I provide myself with the necessary equipment while she evades me by running under the table, diving between chair legs, niftily escaping. Eventually, though, I corner her, grab her and tuck her under my arm and sand-paper or emery-board the over-extending portion of her beak. She squawks, struggles, tries to peck me, grunting protest. When it's done I let her go and she shakes wildly, remaining angry and resentful until she has carried out a ritual session of purification bathing, after which she returns to stand on my knee, talking to me again, ulti-mately forgiving.

I don't know if this beak-growing is common to all rooks, or if, in the wild, constant foraging for food has the effect of keeping the beak in good condition, although in his book *The Crows*, Franklin Coombs suggests that bill abnormalities are common in corvids, par-ticularly the over-extension of one or other mandible, making it difficult for the bird to eat normally or maintain its feathers ade-quately. Without filing, Chicken's beak would be too long. Might her life have been shortened by the absence of my much-resented nail-file?

Each evening Chicken prepares to roost, a process remarkably sim-ilar to our own pre-bedtime routines (although possibly more thorough). By now, her timing has little to do with the natural hours of darkness and winter or summer; she chooses her own time to prepare for bed, usually around quarter to ten. Her night-time preparation is one of permanent fascination, the way in which ritually, with deepest concentration, she begins her prolonged *toilette*, a process she likes to carry out while standing on top of her house. It's similar to her bathing but carried out with more concentration. The cleaning is exacting, beak on feather, in feather, over feather, under feather, inside the interstices of wings, the join of wing to body, the length of the tail, legs, then feet. *Click, click, click,* beak among toenails, down the length of legs, nibble, click, nibble, click. Her tail gets special attention: ferocious diggings, the running of feathers through the beak time and again, each session ending with a swift, vibrating, horizontal shaking of the tail feathers, followed by a quick vertical shake, a fanning – reminiscent of a card player's dextrous shuffling of a pack of cards – accompanied by sounds of clacking and whirring and grunting. There are sounds more of fabric than feather, rustlings and cracklings and snappings, summoning visions of deep crinkled taffeta in the hand, silk whirled through the air, bolts of satin, a Fortuny world of texture, richness, elegance. On finishing her preparations with a series of profound shakes, she settles herself on her top branch and tucks her head underneath her wing. After a short while, she takes her head from under her wing to glare at the assembled disturbers of her peace. Since this takes place in the room in which visitors are most often entertained, her preparations can be embarrassing, a far from subtle intimation that one of us, at least, considers the pleasures of the evening to be exhausted.

cleansing fervour

(Over time, people apart from us have become used to Chicken. Most visiting friends accept being observed, walked around, being deemed acceptable according to some criterion I don't know, after which they're wooed for attention. Some don't mind in the least having trouser legs tugged, or finding a rook launching itself onto their foot or knee as we all sit chatting, although some regard this with more equanimity than others.)

Most people who see Chicken, or those with whom I talk about birds, are interested in their defecatory habits. 'Can you house-train them?' they ask. The only answer is that I don't yet know. If you can, I've failed. But, I reply (in a display of rampant self-justification), there are worse things, many of them found on the pavements of our major cities. Bird excrement is, or at least seems, innocuous when swiftly dealt with (as it is, by me, in a moment's bending and wiping which by now is almost automatic). In quantity, as produced by the doves, it's useful, although powerfully strong, and has to be administered with caution when shovelled into the garden soil. (With enormous generosity, I hand it out in measured amounts to the select among my friends who are gardeners.) As for the general question of the hygiene of my household, which I realise is always to the fore, I just have to work a little harder. I'm far from certain what the repercussions of my neglect might be, but all the same I steam floors and carpets and whatever else I can steam, wash cushions, rugs, clothes, anything that requires it, frequently. In *King Solomon's Ring*, Konrad Lorenz writes of a visit by twenty-four greylag geese, the subjects of much of his work, to his father's study. His elderly father, who was fond of the geese and unwilling to discourage them, was found drinking tea and reading his newspaper while feeding the geese pieces of bread and butter. The effect of their visit upon the beautiful Persian carpet, he says, was still evident, though faded, eleven years later. (He writes too of the parrot, a Blumenau's parakeet, owned by his colleague, the scientist Karl von Frisch, who learned that only by evacuating his bowels would he be allowed out of his cage, a feat he accomplished, sometimes only with considerable effort, ensuring that for a short time at least the Herr Professor's furniture was safe from attack.)

As for more serious concerns, infections, diseases, I try to reassure myself, as well as anyone else who asks. Chicken does not mix with *hoi polloi* of the avian world, the outside birds who might carry unknown diseases or infestations. Since the emergence of the threat of H5N1, I'm particularly careful to observe what's now called bio-security in my dealings with doves, outside birds and bird feeders, a procedure that involves much hand-washing, welly-scouring and the liberal involvement of disinfectant, measures more psychological, I'm sure, than either preventative or efficacious.

For a time, I recall, Saturday mornings were devoted in their entirety to maintaining the hygiene of an unfeasibly large number of creatures. First, the cleaning of the homes of two sets of rats, male and female, five or six or so of each, which involved moving them to temporary accommodation, being careful to keep the sexes well separated, then hefting their large glass houses onto the worktop, washing them out, scraping out the detritus, drying, replacing bedding and rats, replenishing food dishes. That was just the first. Over the years, the numbers grew, fluctuated, fell. In a small check-list of the past, I tick off Miffy the rabbit, Max the starling, Bardie, Icarus, Marley, the doves, Joe and Rosie the canaries, and Chicken. Although it's easier now, with a smaller number to clean, it can be neither ignored nor put off. Often, the anticipation is worse than the task itself. On cold days of winter, or days of horizontal rain when the light isn't sufficient to illuminate the corners of the dove-house, I con-sider the prospect of cleaning them (which I do at least twice a week) with a certain lack of enthusiasm but I've learned that being properly dressed is the thing, welly-booted, fleeced, padded, gloved, when it becomes an odd pleasure, when I relish the balm of it, the rare delight of the bucolic, a faint nostalgic tinge at the memory of the intermit-tent physical work of my youth, peach- and melon-picking, scrubbing vast areas of communal dining-room floors. Now there's a pleasing dissonance, sloshing and brushing as I listen to the sounds of the city beyond the garden walls, the traffic on the Queen's Road, voices from the car park of the Red Cross building behind us. I hurl buckets of hot, soapy, disinfectanty water across the floor, sweeping it before me, swirling it with a dustpan back into the bucket, drying and laying fresh newspaper on the floor, a practice probably deeply at odds with the dictates of the 'proper' bird-house. This has all been made more difficult since the unfortunate advent of tabloid-format newspapers. Even the *Guardian's* slightly larger 'Berliner' format is too small. Did anyone think of the hard-working animal keepers of the nation when they so selfishly redesigned their newspapers?

The cleaning of birds is levelling, as in Mao's China, when the aeti-olated thinkers, the university graduates, the professors and doctors were sent off on shit-shovelling duties in an attempt to reunite them with the land and remind them of the dangers of intellectual pride. It prevents me from imagining that I am anything other than a birds' domestic help.

For all my failures in house-training, I suspect that it's impossible anyway because, with corvids at least, there seems to be a connection between self-expression and defecation, for addressing Chicken in a particularly interlocutory way will bring about an answering, head-bending calling, followed by the luscious, liquid sound, by the *squeeeeak! splat!* that seems its natural adjunct. Spike too engaged in this colloquy of answer and response, and the parrots to a lesser extent – the action, more, far more than just the evacuation of the bowels, appears to be more even than a simple mode of reply, it seems to be an assertion of self. To try to train it out of them might deny them a basic avian right, the right of discourse and opinion. As for keeping a houseful of opinionated incontinents, if their immediate family don't mind, why should anyone else? If I fail to notice a powdery, whitish streak on my skirt or jeans, so what?

(One learns unusual things. I can for instance differentiate, by scent alone, between a magpie, a crow and a rook. Were I to be blindfolded, presented with one of each, *à la* Confrérie des Chevaliers du Tastevin, I would know. Oh random, transcendent gift!)

Now, I'm quite used to sitting of an evening with a rook on my knee. If she's in the mood, Chicken will hop from my knee to the back of the sofa and thence to my shoulder. We sit together under the light, and I feel that she reads as I read, almost inclined to ask if she's finished the page before I turn it. Sometimes, if she's on my knee dozing, she'll relax her objection to hands just enough to enjoy having the feathers of her neck allopreened, the process by which birds groom one another's feathers, possibly for cleaning purposes or pair-bonding or both. I scratch and tug gently at the feathers of her head and neck, and stroke her head. She will, on occasion, return the gesture, standing behind me on the arm of the sofa and digging her beak into my hair. She's gentle but with hair perhaps it's difficult to judge. I want to be polite and so wince as little as I can. I read Dr Lawrence Kilham's description in *The American Crow and the Common Raven* of his allowing his pet raven to allopreen his eyebrows. He says that, in spite of the power of the bills of crows and ravens, they are inhibited about using them 'when performing pair- or social bonding rituals'. I'm happy to take his word for it.

I know that the bodily aspect of corvids, the black, the acute angles, doesn't suggest physical warmth, the tactile pleasures offered by fur, but the perception is incorrect. I dig my fingers down, down into the feathers of her head. There, below the soft depths of her light-grey under feathers, her pink skin, her rounded skull. The deep black feathers at the back of her neck, a warm, rustling collar, provide the perfect spot in which to bury my nose. She groans slightly and makes her creaking sound. She appears to approve. (If she didn't, she would make it clear. She would turn her head awesomely quickly to snap her beak at me.) Her scent is of yeast, of new wool carpet, of warmth and dryness and feather. Rook-sniffing. A strange pastime. Asked once by Konrad Lorenz's wife about the explanation for her husband's love of geese, a psychiatrist friend replied, 'It's a perversion, same as any other.'

9

If Men Had Wings . . .

One end of the mantelpiece in my study is known, in tribute to the banalities of a long-ago Scottish primary education, as 'the nature table'. On it is an assortment of things bought, collected, given, found, many (but not all) bird-related: part of a duck's skull, the empty shells of some small birds' eggs, a rabbit's skull, a few small leg bones and a collection of vertebrae – whose, I do not know – a rabbit's tail. There is a Japanese dish filled with moss out of which sprouts a fan of black feathers, ones we pick up from the floor as Chicken moults, the ones I don't use as bookmarks, and two pieces of quartzite from a Lochaber hill. There is, too, a small collection of raven pellets and a rook's skull.

Birds' skulls fascinate. Hugh McDiarmid's beautiful poem 'Perfect' describes the pigeon's skull he found on the machair on South Uist, the perfection of its emptiness, the absence of the brain that once would have animated bird and wing. The rook's skull was found years ago beside the North Esk River, a smooth shell of grey and parchment bone, arches of eye socket and nostril; it's like a stone under sun, bleached, the long top beak still sheathed in its rhampotheca, the thin shield of keratin that covers birds' beaks. I can slip it from the bone intact, still beak-shaped although fragile. It looks burnished, like fine-shaved metal, like watered bronze. Underneath the skull, a tiny fretwork of interconnecting spurs and spans and broadening bridges. Matching it, I have still, between the pages of a book, the diagram David drew for me after we had found the skull, to explain what everything did and where everything went: nerves, foramena, zygo-matic arch, middle ear cavity, the place where the optic and oculomotor cranial nerves threaded to work what this bird once was, all marked and labelled, foramen magnum, orbital cavity, foramena for the cranial nerves IX, X, XI, XII; *the cranial nerves*, he wrote, *which work the tongue, voice, swallowing.*

The central nervous systems of all vertebrates are similar in their patterns. Our brains and those of birds, of long-ago common origin, are different in organisation although more similar than once was thought. Many functions appear to be analogous. Bird and human brains look different, however. The surface of a bird's brain is smooth, a human's folded and fissured, layered and convoluted with gyri and sulci. Like humans, birds have what they need for what they do – they have a large forebrain, a cerebellum which controls the muscle activity necessary for flight, large optic lobes for the most important of their functions, sight – and it seems that, whilst the structures may be different, the neural connections that allow complex behaviour may work similarly, even when not ordered precisely as ours. Parts of birds' brains have capacities that ours lack, the capacity for neurogenesis, the renewal of cells; hormonally induced seasonal alterations in size of the hippocampus, the area of the brain involved, in both human and bird, with memory, which, affected by levels of melatonin, expands in autumn when finding and hiding food may become important for winter survival. Some birds have, like us, a high brain-to-body ratio, their brains, like ours, larger than is required for the size of their bodies, what is called the encephalisation quotient, a high brain volume suggesting a commensurate level of abilities and intelligence.

Talking of the intelligence of birds, indeed of any species other than our own, even of our own, is challenging. The concept of the intelligence of humans too is one that shifts and changes, is subject to pressures, historical, social, political. It's difficult to know what is meant, what is being measured, when we try to quantify intelligence. Knowing relative brain size, the existence of a cerebral cortex, a cerebrum, an amygdala, tells us only some of what we need to know. Human beings may well have the equipment necessary, the correct weight of matter, the correct number of neurones, axons, dendrites, synapses, all flashing and crackling and sending their illimitable millions of connections, messages, signals, to more or less the desired destinations; the appropriate lobes, all placed approximately at least, in the correct relation to one other; but, as experience has demonstrated historically and demonstrates daily, hourly, this is no guarantee of the subsequent manifestation of behaviour that may be interpreted, even by our own species, as intelligent.

That birds are both intelligent and capable of complex behaviour is clear to those who carry out research into the subject and indeed to anyone involved in any way with birds, but announcements of research findings in avian intelligence are often greeted with amazement. Reports of tool-using New Hebridean crows or counting parrots are regarded as surprising, if not revelatory, for until recently the common belief was that the brains of birds were too small to allow 'intelligent' behaviour.

'Bird-brain' has long been a term used to suggest limited intellectual capacity, but the view of avian intelligence it implies is both reductive and incorrect. The names ascribed during the latter years of the nineteenth century by the German neuro-anatomist Ludwig Edinger to parts of the avian brain contributed in some measure to the widespread acceptance of the idea that the brains of birds were too small and primitive to permit the development of intelligence. His system of naming, relying on an Aristotelian view of the natural world, the *scala natura* which ranked creatures according to 'the degree to which animals are infected with potentiality', combined with a straightforwardly linear notion of evolution by which fish and reptiles were at the bottom of the scale and humans at the top, was to colour man's view of birds and their abilities for the next century at least. Edinger's assumption was that most parts of the avian brain had developed from the basal ganglia or striatum, a brain area known to produce instinctive behaviour, whilst the human brain, being derived from more advanced structures of the pallium, was capable of higher thought; accordingly, he attributed names to the different regions of birds' brains that reinforced ideas of their primordial dimness: archistriatum, paleostriatum augmentatum, paleostriatum primitivum, thereby limiting the opportunities for elucidating or describing any neuro-anatomical basis for complex and sophisticated behaviour in birds. Edinger's designations of birds' brains predominated, more or less unquestioned, until very recently, when increasing research interest began to cast doubt on the reliability of a system that, it became obvious, was both inadequate and incorrect. Scientific studies of the physical basis of bird intelligence and concomitant advances in research techniques into brain function and molecular structures have led to a realisation of the true extent of the prejudicial nature of Edinger's work, and recognition of the limitations it has imposed.

During the 1960s, on retiring from his clinical post at Massachusetts General Hospital, neuropsychiatrist Stanley Cobb began researching avian neuro-anatomy, discovering for the first time that forebrain areas of birds' brains are analogous to the human cortex. In the last few years, a group of international neuroscientists, named with pleasing exactitude the Avian Brain Nomenclature Consortium, have worked together in an extraordinary endeavour, painstakingly renaming the component parts of the brains of birds, relating function to evolutionary development. By so doing, they have elucidated and underlined the fact that the brains of birds are comparable in important respects with the brains of mammals, including humans. A long, detailed list of names demonstrates the changes; among many, the archistriatum has now become the arcopallium, and the paleostriatum primitivum the globus pallidus.

One of the leading proponents of this undertaking is the neuro-biologist Erich Jarvis of Duke University, who says of the new system of naming: 'this nomeclature will help people understand that evolution has created more than one way to generate complex behaviour – the mammal way and the bird way. And they're comparable to one another. In fact, some birds have evolved cognitive abilities that are far more complex than in many mammals.'

In making his famous comment, 'If men had wings and black feathers, few of them would be clever enough to be crows,' Henry Ward Beecher, the nineteenth-century American clergyman, wit and, like his sister Harriet Beecher Stowe (writer of *Uncle Tom's Cabin*), a staunch abolitionist, showed remarkable, indeed admirable prescience.

I have come to recognise too, the dangers of the subject of the intelligence of birds. For me, a chance and random happener-upon birds and their brains, trying to steer a median course between two schools of thought concerning birds and animals, what they think and what they do, is hazardous. I reflect on both. The proponents of one school, the behaviourists, accept only what can be reproduced when it comes to making judgements on animal behaviour, on the results of tests, rejecting, most often, any scientifically unproven suggestion that animals think or feel, or are capable of emotion, eschewing any feeble and unscientific recourse to their personal *anthema maranatha*, anthropomorphism. The other school, the ethologists and cognitive ethologists, are prepared to draw conclusions from studies and observation, to allow what is to behaviourists the unthinkable: conclusions drawn from anecdote, from personal observation, from what just seems to be so, those things regarded as anthropomorphic, loose and gossipy, unrigorous and possibly heretical.

The great American corvid observer Dr Lawrence Kilham writes: 'I have said to myself on a few occasions when watching a crow at close range, "That crow is thinking." The selective advantages of crows forming images and thinking consciously are too great to be dismissed by the dodge of anthropomorphism.'

Anthropomorphism, the ascribing of human characteristics to non-humans, is, at its worst, reductive, a close and pernicious relative of sentimentality, or else it's self-justifying, the desire to mould ideas and images of animals in ways which allow humans to control or understand only within the narrow framework of what we ourselves may be, stemming perhaps from a desire to idealise, or from the atavistic fear of the other, the wish to order the natural world in a way that will reinforce or reflect a morality we may ourselves have lost or forfeited. The words that attribute the more meritorious of human qualities to animals mislead, but the ones that do the opposite, the ones that employ the vocabulary of judgement and superstition, the ones that call birds and animals cunning, cruel, evil, dirty, are the more harmful in both impulse and effect. At the same time, we are too ready to judge by appearance, to accept cultural determinations often made long ago by those who knew little about the animals they were portraying, sometimes incorrectly, as non-aggressive, benign, entirely other than they were.

The American writer and poet Mary Oliver, in her book *Blue Pastures*, sums up in a short and glowing essay the differences between the man-made and the natural, observing that by describing the natural world in the language of diminution, using words such as 'cute', 'charming', 'adorable', we give ourselves power, the power of parents and governors, we are complicit in seeing the world as a place either where we play or a place where we study its other, non-human inhabitants at our will.

But trying to understand, for humans, is only possible through the sole means we have, the filter of ourselves and our fears, our prejudices and often irrational beliefs. The lines we draw between sentiment and rigour are fine indeed. The evolutionary biologist and psychologist Gordon Burghardt puts forward the concept of 'critical anthropomorphism', a concept that encompasses a wide range of critical approaches, physiological, sensory and ecological, which he uses in his studies of animals and their behaviour.

In the foreword to her book *Under the Sea Wind*, first published in 1941, Rachel Carson explains why she has chosen to describe the lives of sea creatures, fish and sea birds in terms that might be considered at variance with the demands of scientific writing: 'we must not depart too far from analogy with human conduct if a fish, shrimp, comb jelly, or bird is to seem real to us – as real a living creature as he actually is'. For her, a tone of empathetic intimacy is a more powerful way of conveying to her readers the mystery of the world of which she writes than would be possible in the language of science with all its cool and measured distance.

In *Mind of the Raven* Bernd Heinrich describes an incident between a human, a raven and a cougar. A woman working beside her cabin in Colorado had her attention drawn by the repeated frenzied calls of a raven flying overhead. Only when it finally landed on rocks nearby did she raise her head to see the cougar that was about to attack her. Her own interpretation of the incident was that the raven, by alerting her, had saved her life. Heinrich's is that, far from warning her, the raven was alerting the cougar to a possible food source, for the likely benefit of both. The raven was behaving as ravens do, appropriately. The fact that some might prefer the raven's motives to be different suggests only that they wish the natural world to be other than it is, as if it might be possible, sensibly, to hope that a raven, for some unknown supra-moral reason, might elevate the interests of a human over its own, or those, incidentally, of a cougar.

I tell people constantly about the intelligence of corvids. I agree when anyone says that they've heard corvids are meant to be intelligent. I've read, observed, believe I know, but do I? How do I know? I haven't done any tests that might prove it, but then I haven't on most people either (or, in fact, on anyone at all), but this doesn't prevent me from forming opinions on the nature or capacity of their intelligence. What I think about people, a matter equally untested, is based on a set of undefined criteria which I recognise but may be different from others pondering the same questions. There is, of course, a comparative basis to any belief about intelligence as much for humans as for birds. I think, much as I might of people, that certain birds I know are intelligent, and some less so. I think the corvids I have encountered are intelligent but how do I know? I base my view on many things: what they do, how they respond, their ability to recognise, their ways of behaving, but on other things too, things untested and possibly untestable, demeanour, curiosity, response. When I read in *In the Company of Crows and Ravens* that magpies are able to recognise themselves in a mirror, an indication of high degree of self-awareness, I wasn't surprised. When a red light is shone onto the white feathers of a magpie as it looks at itself in a mirror, it will begin to try to remove the stain of red. The higher primates and elephants have also demonstrated this quality of consciousness of self.

Some birds appear to be more intelligent than others. Many of the conclusions about comparative bird intelligence have been drawn from observation. Dr Louis Lefebvre from McGill University has designed a system of testing bird intelligence by relating it to reports of innovative behaviour, much of it feeding behaviour, collated from observations made over decades by individuals or published in scientific journals. Lefebvre's method avoids some of the known difficulties of intelligence studies: cultural bias, as well as the use of equipment (which is problematic in studying bird intelligence, because, as many researchers have pointed out, equipment and testing devices of one sort and another are what people do, not birds). According to Lefebvre's study, the cleverest birds (need I say) are corvids, followed by falcons, hawks and woodpeckers. (The question of comparative intelligence may be invidious. Does it matter if one bird or beast is more intelligent than another? The wider significance, or usefulness, may be in the alteration of attitudes, if greater awareness of the capacities of all birds and animals leads to more considered treatment of them.)

There's abundant evidence to show that for a long time and in many different cultures people have recognised, or at least suspected, that corvids are clever, but it's only now, when close scientific examination of the physical properties of corvid brains and extensive study of their behaviour can be carried out, that it's possible to gain appropriate insight into the true magnitude of their abilities.

Corvids are among the birds that have the highest encephalization, or brain:body ratio. Their nidopallium, the avian brain structure that fulfils broadly the same functions as the mammalian neocortex, that of sensory processing, is larger in the corvid than in most other birds except parrots, the other avian group deemed by legend and hearsay to be particularly clever. Corvid brains are relatively the same size as those of apes and indeed Dr Nathan J. Emery, a neuropsychologist in Cambridge's Department of Zoology, suggests that in their cognitive ability corvids rival the great apes and might well be considered 'feathered apes', many aspects of both the biology and behaviour of these two disparate species, in spite of differences in their brain structures, being remarkably similar as a result of the processes of 'convergent evolution' (whereby unrelated organisms, affected by the same environmental circumstances, evolve similar traits). Corvids, like apes, have evolved within large social groups, which by virtue of their complexity and their requirements, their demand for mutual recognition, negotiation and communication, seem to encourage and stimulate intellectual development.

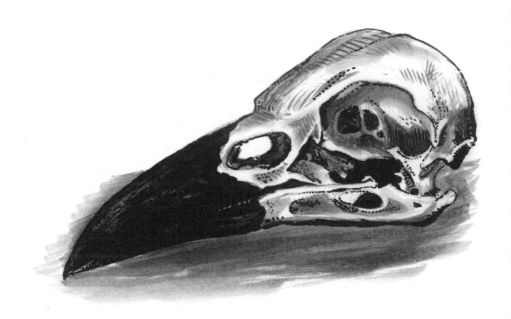

a smooth shell of grey and parchment bone

As a result of their extensive work with both, Dr Emery and his colleague Dr Nicky Clayton suggest that apes and corvids share a number of abilities that contribute to their high level of cognition, including the understanding of cause and effect, and the ability to think about things that aren't present, to apply previous learning to new situations, and to think about the future.

One aspect of behaviour shared by all the birds suggested by Louis Lefebvre's study to be the most intelligent – corvids, falcons, hawks and woodpeckers – is that they are all birds that cache. Caching, which in corvids is part of life in a social group, a bit like lying, requires a prodigious memory, the kind of memory that is accompanied by, and the result of, owning a sizeable brain.

To cache, a bird must be able to remember where its (often unimaginably) large numbers of cache sites are located. It needs the spatial awareness to be able to find its cache sites again. In addition, birds may have to have what is called 'theory of mind', the ability to consider the mental processes of another, which in this case involves being master of a range of stratagems of Byzantine complexity in order to deal with the theft and duplicity that seem to be integral to caching behaviour. Birds have to be observant, vigilant, sneaky, possibly dishonest, and, more than all this, they have to be able to anticipate that their fellow cachers are likely to be just as prone to these unfortunate tendencies as themselves. Stealing from other birds' caches is common. Birds who know that they have been watched in the act of caching will return later when unobserved to remove and rehide their store. Interestingly, it's only those who themselves who have been thieves in the past who will do this, indicating that they have the ability both to remember and to ascribe motive.

Not all birds, of course, are as intelligent as corvids appear to be. From long observation, I've always thought doves and pigeons considerably less bright than the corvids of my acquaintance (and indeed it seems likely that they possess a smaller nidopallium than corvids), but a piece of innovatory research carried out at Keio University in Japan demonstrated that, with suitable training, they can tell the difference between paintings by Picasso or Monet, Braque or Delacroix. The training, I'm sure, can't have been easy, even though food rewards were involved. I have noticed no particular sign of highly developed aesthetic appreciation in my doves but I'm sure that it's only because I have denied them the correct opportunities to display it. (Might an auction house be interested in this singular skill? 'I say, that's a nice Bonnard!' 'You're not trying to tell me that's a Watteau!') For their efforts, the three researchers who carried out the project were awarded the 1995 Ig Nobel prize, a prize awarded annually for research that 'first makes people laugh and then makes them think', by the scientific magazine *Annals of Improbable Research* at a ceremony in Harvard's Sanders Theater.

I don't find it difficult to believe that some birds have the capacity to do what might be described as thinking. Observing in my own birds what seems like a process of consideration, followed by an action, perhaps a choice of one from among a series of possible actions, allows me to believe that it has been thought of one sort or another that has brought about that particular choice. It was the observation of his behaviour that allowed me to think that Spike was intelligent, more intelligent than any of the other birds I have met. When he balanced an object – a pamphlet, a rubber glove, a matchbox – carefully on top of a half-open cupboard door, as he did frequently, watching, waiting until it fell onto the head of the next person to open the cupboard, was that behaviour a result of thought? When he tried to fly with his chopstick, laying it down to check the point of balance before lifting it? Possibly.

We used to see Chicken hiding things in the garden, worms, small stones, pinning a leaf over them with a twig to hide them and keep them in place. I watch Chicken caching. She now knows the texture of things, what will stick to what, what is best cached in a place where stickiness is or is not required. I have seen her smooth out paper so that it will fit between the lathes of the wall. Had I thought of it earlier, I might have tried to teach them Greek and Latin.

I don't want birds to be other than they are. I don't believe that they understand every word I say, the dictum beloved of animal lovers, pet owners everywhere, which means, in fact, that they understand intonation, expression and body language. Chicken understands a great deal – gesture, tone, inflection and, I'm certain, many words – but some words she doesn't, as I may not fully understand some words of hers. By now, I don't worry about it. We make ourselves understood.

According to the behaviourists, it is impossible not only to make judgements about the emotional lives of birds, but to countenance, without strict and sufficient evidence, that they have feelings or emotions at all. The cognitive ethologists are more accommodating. I relate it to my own experience. How else can I interpret frank displays of outrageous fury, the hissing and feather-raising of the enraged bird? What else are the displays of anger and frank dislike Bardie feels for me, and equally, the pleasure, affection, indeed love, he shows for Bec? If it wasn't rage that made Icarus shout in apparent irritation at Bardie and bite his feet, what was it? Why, if we are not all subject to at least some of the same emotions, are they so recognisable? (Do I just wish it to be so?) Why, if we all live in situations of complex social organisation, should we have emotion and they not? If they are capable of anger, might they not be capable of affection? Whilst it may be easy for humans to apply values, insights, moralities that may or may not justifiably be applied to animals, that does not, clearly, necessarily suggest that tenderness, kindness and care for others do not exist in any but human society.

I use the words of human emotion to describe bird emotions, for I know no other ones, only passion and anger, delight and love and grief. There are plenty of incidences of the observation of grief in animals, although judging the mental processes impelling their behaviour is difficult, if not impossible. I have, though, seen what looks like grief in doves, watched their sinking, bowing into postures of abject despondency on the death of mates. It looks remarkably similar to grief in humans. I have seen – or believe I have seen – in birds impatience, frustration, anxiety in the urge to impart news, affection, fear, amusement (the last being a difficult one, I admit, to prove, merely on the basis of watching the look on a magpie's face as its booby-trap was successful) and, particularly, joy. I may be wrong in my interpretation, inexorably skewed in my judgement by the effects of long exposure to anthropomorphic ideas, but I believe that I recognise all of them because they too look remarkably similar to their human equivalent.

Some corvid young remain with their parents, 'helping' to rear the next clutch of young, but 'help' is a word that suggests intent, and for some biologists is therefore a word misapplied. Why should birds 'help' one another when the sole imperative in the lives of birds is meant to be the self-interest of genes, reproduction at any price, any-thing more than the urge to procreate and to survive being superfluous, unlikely or impossible? Crows have been observed engag-ing in behaviour that would have no reproductive value – feeding other ill or injured crows, taking care of orphaned young – and while it may be that such behaviour strengthens social and family bonds, encourages mutual defence and eases the processes of further repro-duction, that may not be as different as we would like to think from our own behaviour and underlying motivations when we engage in similar acts. Studies involving egg-swapping among corvid populations with observably different social customs have demonstrated that co-operative behaviour is learnt, rather than innate. Those who have studied pair-bonding among birds, such as Bernd Heinrich, have con-cluded that birds 'fall in love' (as would seem appropriate enough where, in general, lifelong monogamy is the social norm).

The concept of empathy in non-humans is also questioned. Recently, a vet writing in an American veterinary journal suggested that whilst animals sometimes respond to their owner's distress, their response should not be interpreted as their 'experiencing the emotion of empathy'. Really? I have frequently experienced responses from both Spike and Chicken that I can describe or interpret only as empathetic. I may be wrong, too anxious to see them as capable of such emotional range, but to have a magpie, on seeing me weep, hover on top of the fridge, wings outstretched, tremble for a few moments then fly down to my knee to crouch, squeaking quietly, edging ever nearer until his body was close against mine, seemed to me at the time (as it does now, on reflection) an act of unexpected tenderness that I can interpret only as empathy. Chicken too will seek to be close to anyone in an apparent state of distress, jumping onto their knee, bending her head forward, attempting to be as close to them as possible. There may be other explanations of their behaviour, but I can't, at the moment, think what they might be. The fact that relationships between members of our own species are difficult enough to interpret and understand makes me wary of drawing conclusions from Chicken's behaviour towards Marley, whom she would often visit in whatever room his house had been placed. She'd stand near him for a long time, mostly in silence. I don't know if there was a bond of some nature between them, some avian connection, but I like to imagine that Chicken, in her way, tried to communicate with and reassure that nervous, anxious conure.

It's often suggested that, without language, thought (and indeed consciousness) is, if not impossible, then limited, but is all human thought framed within language? Is there not impression, sense, a series of sensations, underlying the process of thought? Might it not be that there are other ways to think? Do we know enough of another species' language to know?

Certain ways of behaving are known to be the adjunct of intelligence in birds, playing and caching among them. Although some scientists are sceptical of the use of words such as 'play' being applied to animals or birds since it's behaviour that would bestow no advantage in terms of survival or reproduction, many others recognise the extensive play behaviour of animals and birds, describing snow-boarding ravens, crows who drop twigs then fly to catch them before they reach the ground, the teasing, chasing, amazing aerial displays. Why should it be different from the benefits of learning, social cohesion and enjoyment that humans derive from play? When I say that Spike enjoyed playing, I can think of few reasons that would have made him engage so vigorously in his magpie version of football, on his own or with anyone who was prepared to engage in this unusual sport, chasing a ball he had nudged with his beak, running after it when it was kicked for him by someone else, other than simply for pleasure. He loved play-fighting of all sorts, and his collection of toys, which he put carefully into and took out of his toybox.

Recently, coming out of work, David saw in the grounds of the hospital three rooks, this year's young, who had found a discarded but not empty lager can. They were pushing the can to and fro, jumping on it, rolling on the ground, trying to drink the remains of the lager. (It was after rain. There was plenty of water in puddles to satisfy thirst.)

It may of course have been, not play, but 'brand recognition', like that of the crows described in *In the Company of Crows and Ravens* who have learned to recognise the logo of a well-known fast food company; perhaps the Scottish rook, so habituated to the sight of this particular lager, has decided that no other will do.

Chicken loves to play with paper, ripping it, tossing it into the air, scattering it. As I write, she's playing enthusiastically with an elastic band. For years, she played with the rubber mice I bought for her, throwing them around, pouncing on them, carrying them by the tail in her beak. Blue and red and yellow, they were routinely, inadvertently thrown out by me during house-cleaning. More were bought to replace them until the day when, very suddenly, for a reason I don't know, rubber mice disappeared. Someone somewhere had stopped manufacturing them. Everywhere I go now, every foreign city, I seek out pet shops, but without joy. The sole world source of rubber mice has gone. Instead, I've given her a rubber frog which she treats with much the same vigour as she did the mice although the frog is rigid, lacks the kind of tail by which it can be given a punishing shake. She doesn't think they're real, either frog or mouse. She doesn't like real mice. In autumn, when one or two intrepid field mice find their way into the house, I watch Chicken's reaction as they make their way across the study floor towards the scattering of food lying beside her house, the way she notices instantly that the mouse is there, the nonchalance with which she begins to look away, the casualness with which she jumps from the top of her house onto the floor cushion, the concentration with which she searches for something in one seam or another before being overcome with a sense of urgency as she recalls that she has important though as yet undisclosed business in the kitchen. Later in the evening, I'll find her standing uncertainly on the strut of a dining chair, probably wondering if it's safe to go home. I'm glad she takes this view. I wouldn't like to have to deal with the consequences of territorial aggression. I clear up the food scraps and hope, as Chicken probably does, that the mice will go away.

It's the points of similarity between us that delight me still. I admire the birds' anger and their rage, for I too (perhaps about different things) feel anger and rage. I like seeing their apparently purposeless play, for it indicates to me that they have minds free enough from concern to do it. I am astonished, always, by the way they'll appear to know without knowing, to understand, anticipate, react, for it makes me feel as if I live in an indivisible world, that my belief that we're nearer in every respect than I could have imagined is correct, that we are, whatever we are, something of the same.

10

A Few Ticks of the Clock

Time weathers us, wears away the differences between us. I think more now than I used to about Chicken, about all birds, about myself, about how we got here, how we came to be, and I realise that what makes us the same interests me as much as, more than, what makes us different. I'm fortunate in having this unique opportunity to observe, or to grow accustomed, for without it I'd know the differences, but not the similarities. I might have known about birds from reading, about their distribution and characteristics, the biological facts, the ways one calls them what they are according to the fluid requirements of Linnaean taxonomy: kingdom, phylum, class. I might have known the definitions that order us, place us, explain us, but I wouldn't have known anything else.

Having observed, lived with a corvid for as long as I have has changed things, has reclassified us both, readjusted my observations, my consideration of both our places in the world. I sit with Chicken on my knee in the quiet of evening, both of us in post-working-day philosophic mood, and it's the moment when I follow our progress down through the stages of where we were and where we are, through the words that place us so precisely in evolutionary time. I begin with the fact that we're both vertebrate, of the kingdom animalia both, both of the phylum chordata. I like this. It's comforting. Hey Chicken, we're in this together! Backbones! But then abruptly, alas, at class we part and at that point wander off separately into time and the development of our avian or mammalian ways, into our feather and our fur, into the byways of our biology and the benign or malign idiosyncrasies of what and who we are. But still we carry on down, further and further, into order, into family, into species, and on reaching this, the end point, or the sole end point of which we can have any certainty, looking across these last and final categories, genus, species, as *Corvus frugilegus* and *Homo sapiens*, again I am aware, as I always am, that there's more for me to feel anxiety and guilt about than her.

I often wonder too about the point at which our evolutionary courses diverged, when she ascended on the path that would take her to an arboreal life of black feathers and wings and the enviable attribute of flight, while I proceeded stolidly through time to become what I am. *Don't leave me behind*, I want to say to her, *I want to come with you*, but there's not much she can do about it, or I for that matter, it's too late, a few million years, 280 million, give or take. Off she goes into the distinguished, elevated society of ravens, jackdaws, the witty and eclectic circle of the crows of the world, while I – well, I find myself in mixed company.

At these moments of ontological reflection, it's reasonable, even appropriate, for us to think of archaeopteryx, the first bird. I read, if not to Chicken then to myself, Edwin Morgan's wonderful poem 'The Archaeopteryx's Song' –

> *I am only half out of this rock of scales.*
> *What good is armour when you want to fly?*
> . . .
>
> *I saw past and present and future*
> *like a dying tyrannosaur*
> *and skimmed it with a hiss.*
> *I will teach my sons and daughters to live*
> *On mist and fire and fly to the stars*

– and as I read I wonder, is Chicken, her grey rook's face as familiar, as dear to me as would be that of any dog, any cat, representative of those who left behind the rock of scales, the weight of stone and armour? I look at her. Was she once a dinosaur? A small dinosaur, a maniraptor, but a dinosaur none the less? The answer to both questions seems to be yes. The qualification is there only because disagreement about the lineage of birds continues, a kind of three-sided contest with the fossil evidence at its core, to be disputed, taken apart; the argument of Linnaean classification versus cladistics or phylogenetic systematics, the argument of what, in fact, a bird is and how long it has been one anyway; everything dependent on how the facts are assembled, on taxonomy or on belief, who belongs where, and for how long they've belonged there, arguments of definition versus diagnosis. (Definition tells you who is in a group. Diagnosis decides whether they should be there or not.) Once, if you had feathers you were a bird, but now, since the discovery of feathered, non-flying dinosaurs, even feathers don't guarantee you any confidence in yourself and your place in the scheme of evolution.

The Linnaean system of classification, binomial nomenclature, building on Aristotle's work two thousand years before, systematised the living world, made it possible to identify each organism, whether orchid or lemur, moss or crow, according to a hierarchy of kingdom, phylum, class. Wonderfully comprehensive as it is, it doesn't take account of the evolutionary antecedents of any given organism. According to Linnaean taxonomy, all birds are of the class aves. Since nothing else is known to have feathers, feathers are a definition of 'bird' – if it has feathers, it's a bird, and if it's a bird, it has feathers. But, according to phylogenetic systematics, a clade – a single group of organisms – constitutes one ancestor and all its descendants, and since birds are (according to the view supported by cladistic analysis) descended from dromaeosaurid theropod dinosaurs, they are, in cladistic terms, dinosaurs.

There are other pressing questions, questions of relationships, questions of flight. Who flew and who didn't? If they did, how did they? How flight evolved is still uncertain, since functional morphology – the study of the relationship between physical form and function – has not yet resolved one of the major questions about flight, whether it had its origins in the movement upwards from the ground, with creatures (ungainly, clumsy creatures) running, then leaping, then taking to the air, the cursorial theory, or in their movement downwards, of leaping from tree to tree, a movement that became a pre-flight gliding, the arboreal theory. While the latter is the more likely – gliding down from a tree being rather easier than the alternative, which would have demanded considerable force to work against gravity – dispute remains. The 'insect net' theory, presented by the palaeontologist John Ostrom (an adherent of the cursorial theory), suggests that using forearms in a forward movement to trap and catch insects was the precursor to their development into wings as tools of flight. And all the time, the insistent voice to the side is shouting, '"*In the beginning . . .*"!'

Tracking through the ages, through the unimaginable lengths of palaeontological time, 'deep time', negotiating those three mysterious letters that describe aeons, MYA – 'millions of years ago' – through Jurassic, Cretaceous, Paleocene, Eocene, Oligocene, Miocene, it's possible to stalk the progression, trace the slow appearance of the relatives of the birds we know, and backwards too, from the birds around us to the first fossil specimen ever found that might have been a bird, archaeopteryx.

Archaeopteryx, tricksy, contentious, beautiful fossil of the beautiful name, *Archaeopteryx lithographica*, 'ancient wings from the printing stone'. The first specimen of archaeopteryx was found in 1861, a spread of elegant bones, neck arched backwards by the tightening bow of death, feathered wings preserved in finest detail by the carbonate limestone of Solnhofen in southern Germany. This *Urvogel,* protobird, emerged into the excoriating light of the modern age after 150 million years of darkness, of sharing the sediments of shallow, subtropical Jurassic lagoons with plant remains, with pterosaurs and insects, the forerunners of those who live today – ephemoptera, hemiptera, coleoptera – and with medusae, ammonites, crinoids and all the other palaeospecies held in states of near-perfect preservation by the anoxic conditions of the lagoon water, protected by the limestone deposits that would one day, because of their porous smoothness, be the medium of a different creation, becoming the lithographic blocks used by Degas, Munch, Escher, Toulouse Lautrec. A number of specimens have been discovered over the years but the best-known is the Berlin specimen, so called because it was bought by Humboldt University's Natural History Museum, where it still resides safely in a heavy vault, providing as ready a source of dispute and minute examination as it did when it emerged into the light and heat of the arguments surrounding the nature of creation in the years after the first publication of Darwin's *On the Origin of Species* in 1859.

Archaeopteryx's fame comes from its significance in providing what, to some at least, is the incontrovertible link between dinosaur and bird. He remains until now the oldest specimen of feathered creature ever discovered, unique not only in his age but in the fact that he was the first fossil found to combine features of both reptile and bird. Archaeopteryx has a small skull with widely spaced, backwards-pointing teeth set into narrow jaws, clawed fingers, a long tail of twenty-two unfused vertebrae, all of which are reptile features; but he has feathers too – asymmetrical flight feathers, leg and back feathers – and bones that are hollow like the bones of the birds around us today. He has anisodactylic feet (three front toes, one back, like modern passerines) with the hallux, the first toe, opposable; a furcula, or fused clavicles (what on our plate we'd call a wishbone); and a long, bony tail. He may have been ectothermic, cold-blooded, like reptiles, or homeothermic, warm-blooded, like modern birds.

A CT scan of archaeopteryx's skull and subsequent 3D reconstruction of his brain, carried out in a collaboration between the British Museum and the University of Texas, show that his brain was remarkably like that of a modern bird, although smaller, and that he had brain and inner-ear structures that would have allowed him to see and hear well, to balance and to fly. (I try to imagine what it must have been like to see the detail, the imprint of blood vessels, the space where the brain of a creature 150 million years old once was.) He probably had the experience of flight but it's not certain what kind. While his brain would have allowed him to glide, the limited development of his shoulder bones might have prevented flapping flight.

The fossil record for birds is sparse by comparison with other creatures, the lumbering dinosaurs, the armoured, plated, plaqued boneheads Edwin Morgan writes of in his poem. Birds were, as they still are, light, insubstantial, hollow-boned things, lacking the weight that would have anchored them firmly, solidly in the mud to lie until chipped out, bit by bit, hacked or bulldozed, picked or brushed from the earth that was for millions of years their tomb and the source of their preservation. Birds must have been, as they often are in death, blown away, trampled on, left to fold invisibly into earth, swallowed into the seething, busy depths of Jurassic seas. I think of them when I find small corpses occasionally in the garden, damp-feathered, camouflaged against soil, finches, a blackbird with its beak dulled into near-invisibility, the dead blue-tit lying on the stones of the path, colourless, now almost unidentifiable as what he was, even after a brief few days reverting back, melting into the substance of the earth.

(Fossil brains, like fossil birds, are rare, soft brain tissue decaying too easily for preservation except in very specialised environments. CT imaging and 3D modelling are the newer methods of reconstructing ancient brains, but most of what's known about brain evolution has been learned by the use of 'endocasts', model brains produced using the cranial cavity as a mould. It was Tilly Edinger, daughter of the neuro-anatomist Ludwig, who, inspired by her love of childhood visits to the Senckenberg Museum in Frankfurt, became interested in fossils, working at the museum until 1938 when she was obliged to leave Germany. Moving first to Britain, she then went to America where, during pioneering years of study and teaching at Harvard, she became a distinguished founder of the field of palaeoneurology, the study of fossil brains.)

Thinking of archaeopteryx, I imagine him to have been big, as big as his reputation perhaps, as weighty as his provenance, his importance, his life among dinosaurs, but he wasn't: he was small, the size of a magpie, the size, I assume, of my own magpie, Spike. One reconstruction of him that I have seen makes him black and white, with a slightly mad look in his eyes and, apart from the rather odd addition of teeth, he looks remarkably, unnervingly like Spike.

If the very existence of archaeopteryx suggests a transitional stage between dinosaur and bird, acceptance of this has been, in the years since its discovery, strongly resisted by all those who would rather believe that birds sprang, or rather flew, fully formed from the hand of the Almighty at some indeterminate moment on the fifth day of His famously busy week, and by those who continue to argue, among other things, that archaeopteryx, being definitely a bird and therefore not representative of any transitional stage, proves conclusively that birds did *not* evolve from dinosaurs.

The world of creationism is more complex than I could have imagined, with 'old earth' and 'young earth' creationists, 'progressive' creationists and 'gap' creationists, as well as the adherents of a wide spread of theories: theistic evolution, modern geocentrism, intelligent design. According to 'young earth' creationists at least, the earth is less than ten thousand years old, a difficult position indeed, since, even beyond the confines of disputed palaeontological proofs, there are clear written and other records on earth of human civilisations that have been in existence well before the given date (apart from everything else, apart from science, apart from the utterly obvious, reliably proven facts that demonstrate that the earth and all that it contains have been around for more than a few thousand years).

Although Christianity dominates in ideas of creationism, it's not the only religion to hold similar beliefs. Creationist wings exist in Islam and Hinduism as in Judaism, although many of the more prominent

ancient wings from the printing stone

Jewish theologians – Maimonides and the Gaon of Vilna among them – believed the Torah to be a theological and not a scientific text and therefore not inimical to ideas of evolution. The Jewish mystic tradition, Kabbalah, embraces ideas of evolution, one theory of creation being that of the sixteenth-century Safed kabbalist Rabbi Isaac Luria, whose idea of *zimzum*, or the contraction of light, in some ways presages future theories about the origins of the universe.

The view of at least one trend in current ultra-orthodox Jewish thinking is that Genesis must be interpreted literally, a fact recently impressed on me by discovering the reluctance of a near relative, member of an Orthodox community, to pass on to her many children the single wall-chart, among dozens given away free by one of the newspapers, that related to dinosaurs. Those depicting stars, planets, clouds, amphibians, spiders, rodents, sheep, wild berries, fungi, goats, trees, farm animals, poultry, gemstones and apples met with no such censorship, being deemed, apparently, safe to be given into the hands of the impressionable young. 'Hey,' I say to the cousin who tells me this about his sister, 'I didn't know we don't believe in evolution.' 'We do,' he says. 'They don't.'

Genesis is beautiful in concept and expression but, as a document upon which to base one's belief in the origins of the universe, imprecise, sketchy on the details. Birds, according to Genesis, were created on the fifth day: '*ve'of yeofef al ha'aretz al p'nai rekei hashamayim*' – 'Flying creatures shall fly on the face of the heavenly sea' – and on the sixth, creatures: '*tozi ha'artez nefesh chaya l'minah behema v'remes v'chayeto artez le minah vayehi-chen'* – 'Earth shall bring forth particular species of living creature, particular species of land animal and beasts of the earth. It happened.'

Thus, the creation of birds preceded the creation of beasts, obliging creationists, in their unwavering adherence to the blankly literal, to spend inordinate amounts of time and energy in elaborate refutations of palaeontological evidence to the contrary, using bad science to disprove good science. Some insist that the evidence is faked, that the world just seems and looks old but isn't, thereby allying the Almighty with a particular type of disreputable antique dealer in making what is a mere few thousand years old look millions of years older.

It's not only fundamentalist belief that has cast doubt on the palaeontological significance of archaeopteryx and its contribution to the understanding of evolution of birds from dinosaurs. The late Nobel Prize-winning astronomer Professor Fred Hoyle and his colleague Dr Chandra Wickramsinghe deemed the fossil specimens themselves to have been forgeries, suggesting, against all available evidence, that feathers had been mischievously and fraudulently added. Since they were also the people who suggested that insects are more intelligent than human beings but choose, by means of an insect-wide conspiracy, to keep it to themselves (in addition to a broad array of theories most tactfully described as 'controversial'), there might be place for some doubt about their claims.

Much is still not yet known about the origins of birds, although the discovery in north China, at Liaoning and Gansu, of feathered dinosaurs, some flightless, has been illuminatory, exciting, for *Microraptor gui*, who lived during the early Cretaceous, 125 million years ago, with its long tail feathers, and flight feathers on its hind legs, is regarded as another transitional creature between small carnivorous dinosaurs and birds, as is *Caudipteryx zoui*, a dinosaur with short down feathers on its body and long tail and wing feathers, a creature denied the possibility of flight by the shortness of its 'wings'.

Confuciosaurus sanctus, a Liaoning resident of long standing who may be of similar age to archaeopteryx (although unrelated), has a pygostyle – the bony fusion to which long tail feathers attach (a parson's nose, by any other name) – large claws and a wing shape not unlike the wings of modern birds. He is a rival for archaeopteryx in terms of world firsts, being the first beaked, toothless bird. He is named, aptly enough, for another elderly, venerable, possibly similarly toothless inhabitant of the Middle Kingdom, the great philosopher Kung Fu-tzu (transliterated into the Latinate 'Confucius'), a man whose ideas on the desirable nature of ordered relationships between people have been sufficiently enduring, history and time notwithstanding, to have exerted a lasting influence on interpersonal and intergenerational relations between Chinese people everywhere.

The discovery at Gansu in 2003–4 of many well-preserved fossil birds dating from the early Cretaceous, subsequently named *Gansus yumenensis*, has been described as the finding of a missing link in bird evolution. The specimens, particularly well preserved by the fine sediments of the ancient lake at Changma, are similar to birds we know today, with at least some of the same physical features, including webbed feet and feathers. *Gansus yumenensis* were probably like small grebes or ducks and have been credited with leading birds through an aquatic phase, to emerge and fly far from their heavy, dinosaurian beginnings.

The wide diversity of bird-life around us began in the south of the earth, in that vast territory known as Gondwanaland. In the branches of beech and conifer and pine forests, amid the mosses and flowers and ferns, the wild richness of flora, neornithes (modern birds, or most of them), have their origins. A hundred million years ago, Gondwanaland, now Australia and New Guinea, was beginning its slow disintegration, a shifting, breaking jigsaw-piece-scattering that transformed the surface of the planet, moving continents, thrusting land areas north, spreading oceans between them, allowing and encouraging the spread of species northwards. The Cretaceous, 65 million years ago, seems to have been a particularly busy time for birds, with species expanding, radiating, evolving into what would today be familiar species sharing the sky with species no longer in existence. (Unlike most animals, some birds seem to have survived the K-T mass extinction event of 65 million years ago, which is believed to have been caused by the collision of a meteor with Earth.) By the early Eocene, 50 million years ago, all major orders of birds existed, and over the following 15 million years most modern families appeared. The Miocene, 23 million years ago, seems to have been boom time for passerines – most of the species with which we are now familiar would have been extant. (How easy it is to talk of millions of years, how hard to stop in this moment and try to expand one's imagination to see the slowness, or the fastness, of time.)

Bio-geographical and phylogenetic studies of corvids and other birds have allowed biologists to trace the movements of birds through time and place, to find their relationships and origins and the genetic close-ness of species now found continents apart. The oldest passerine found in Europe, a hummingbird, dates from the Oligocene, 30 million years ago. Named, originally enough, *Eurotrochilus inexpectus*, 'Unexpected hummingbird' (presumably because of the singularity of its being the first hummingbird fossil discovered outside the Americas), it was found at Frauenweiler in Germany, putting the first appearance of passerines earlier than had previously been thought. Corvids appear from the evi-dence to have been around in one way or another since the Upper Miocene, the time when passerines in particular were expanding, moving from their origins in Gondwanaland as far as North America – where fossil evidence shows their presence during the Pleistocene – and Europe, where one of Chicken's more recent ancestors, the fossil corvid *Miocorvus larteti*, dating from the Middle Miocene period, 17–14 mil-lion years ago, was found in France in 1871, and *Miopica*, the ancestor of magpies, in Ukraine. Corvid fossils in North America date from the Pleistocene, 2 million years ago.

Evidence for the existence of human or pre-human life in Britain suggests that first attempts at settlement were made some seven hun-dred thousand years ago and, whilst further approaches appear to have been made at intervals by Neanderthals and Cro-Magnons, the vagaries of water and ice overcame them all. Continuous human habitation seems to have begun in what is now southern England a mere twelve thousand years ago.

In *Birds and Men*, E. M. Nicholson writes of the first settling of Britain by humans not yet in possession of the implements of development – spade, plough and wheel – of a land, now densely populated, covered by trees and water, and of the 'first long phase of human colonisation', the process by which man's role evolved from hunter to cultivator and farmer. It must, he says, have been jackdaws and crows, 'sharp-eyed opportunists', already here, already long-established, that were the first to adapt to life with humans. They must have been the first birds to witness and indeed experience man's anger in defence of what he regards as his own, the first to lose sovereignty to the ways he chose to live on earth, those ways that have brought us to where we are today. The time since its inception seems so recent, after the airy talk of MYA, the processes of man's effect upon the earth, quick and quickening, so sudden, and accelerating, and I want to freeze the frame, myself and Chicken in our miraculous, chance convergence, the time of which Louis J. Halle writes in *Spring in Washington*, his beautiful, acute observation of spring 1945, of the nature of human existence, of the clock that marks out eternity:

For a few ticks I am here, uncomprehending, attempting to make some record or memorial of this eternal passage, like a traveller taking notes in a strange country through which he is being hurried on a schedule not of his making and for a purpose he does not understand . . .

11

Conversations with a Magpie on the Nature of Consciousness

Sometimes, not often, I play the recording I have of Spike's voice. 'Hello!' it yells, brightly, with assertive confidence. No one who didn't know would think that the voice belonged to anyone other than a person.

These days, I see fewer magpies than when Spike was still here. When he was, they were everywhere. They seemed to surround us, in the tall tree in the neighbour's garden, dipping every day in pairs round the garden pond (no doubt persecuting the unfortunate frog), flying to and fro across the sky with their glittering, lilting flight. Now occasionally I see one or two, standing perhaps on the tall chimney of a building nearby, the place where the doves, when let loose, like to congregate. I greet them, not from superstition but from the desire to communicate again, however distantly, with a magpie.

I think about why I don't see them round the house and wonder frequently if the neighbourhood magpies knew that Spike was here, if an unimagined network of magpie communication spread the news that in that house, *that one there*, one of their own was imprisoned. I might have been spotted catching him. The scene might have been witnessed from high in the prickly branches or a nearby chimney, that undignified chase around the foot of the monkey-puzzle tree, the moment of snatching, the stowing of the small creature in the brown-paper bag. They might have followed the car home, watched me carry out my prize, my tiny prisoner.

I think of all those moments, the one when I reached him under the tree, the one when he spoke, the moment when his word became the definitive decision on his future, when it was obvious indeed that he was ours for life.

Spike settled as Chicken had done, easily, slipping with reassuring calm into his place in the family, becoming one of us, a family member. With no discernible anxiety, he ate and thrived and grew. As with Chicken, I consulted Professor Lint, reading with the fervent anxiety of the new parent, as one might with an infant moving from milk to solid feeding. In deference to his advice (but still eschewing rodents and chicks) I bought mealworms from the pet shop, cautiously untying the knot in the polythene bag, emptying the squirming, tumbling mass into the small dish from where Spike would gobble them with horrifying, rapacious delight, cramming the pouch under his chin to bulging. He'd hunt escapees relentlessly across the tiles of the rat-room floor, snapping them in one by one, where they'd wave and squirm between the rictus bristles at the sides of his beak.

It was Spike's perspicacity that I began to notice first, the piercing quality of his observation, the sense I had that among us was a stern beholder. Almost immediately, he asserted his own place in the family order, and while I can't say that he responded to parental discipline, he appeared to know that in theory he should. He showed a certain juvenile's respect for David and me, whilst his relationship with Han was from the beginning utterly different, that of a combative sibling, with whom it was possible, if not desirable, to fight and banter and swear. By the time he came to us, Bec had already left home. He seemed always to be uncertain with her, watching her vigilantly, knowing that she was part of the family but unsure about precisely what his own relationship with her might be. Eventually, he decided that she occupied a place between Han and me, a sibling but one who should be treated with a degree of measured respect.

As he grew, his running, climbing, clambering, scaling, swarming became more confident. There was no point in doing as we had done with Chicken and clipping his flight feathers, since he could climb as easily as he could fly. He established his own flight routes, his stopping places, areas of interest, floor to chair-back, chair-back to fridge, reaching in an instant his favourite place, the top of the kitchen cupboards, his personal realm. (We disconnected the cooker hood. The consequences of a meeting between a powerful air extractor and a magpie were not to be contemplated. Spike employed the unused cavity as a cache site.) He would run up the flights of stairs to the top floor, poise himself on the edge of the banister then launch himself into the deep space of the stairwell to spiral downwards to the ground floor, where the breathless pursuer would find him already taking his ease, preening his feathers casually on the kitchen mantelpiece. If I left the kitchen door open, I'd find him in one of the girls' rooms, investigating. I began to be more concerned about open windows.

As I had with Chicken, I worried about Spike, knowing that the same Faustian pact would carry the same uneasy guilt. The bargain, though, was the same. In the wild, magpies live short lives; short, although brisk, busy, vital and pugnacious lives. The disparity between the longevity of birds in captivity and those in the wild is remarkable. According to Kenton C. Lint, 'Magpies are long-lived . . . reaching ages of fifteen to twenty years in well-planted aviaries.' In the wild, they fare less well. One study, quoted in Candace Savage's *Bird Brains*, suggests that a fifth of fledgling magpies are killed by predators within two weeks of leaving the nest. Most live for two and a half years, or less. It's unlikely that Spike would have survived had the cat arrived before me. I reminded myself again of the difficulties of reintroducing a bird to the wild. Again I told myself that the choice had been not between this life and freedom, but this life and death.

In everything, Spike lacked Chicken's innate caution. Suddenly I understood those photographs in the books, of magpies squaring up to buffaloes, eagles, wolves. He was without fear or inhibition. His curiosity was terrifying. As an infant (quite a large one) he squeezed himself one day into Bardie's empty house. Bec, at home for a holiday, had taken Bardie upstairs, as she used to do in the days of their youth when they'd spend hours together in her room at the top of the house. I watched in amazement as Spike, even then considerably larger than a cockatiel, forced himself through the small door of Bardie's house and, hunching, folding himself small, managing to perch awkwardly, like Alice in Wonderland in the White Rabbit's house after she has drunk from the bottle marked DRINK ME, on one of Bardie's perches. His tail stuck awkwardly from between the bars ('Sure, it's an arm, yer honour!'). As soon as he noticed me looking at him, before I could call everyone else in the house to see, he shot out, if one can carry out so swift an action while squeezing through a door much too small for one's size, and flew to his lair above the kitchen cupboards ('Catch him, you by the hedge!'). I can't think of any reason he might have done it apart from curiosity, a desire to see the world from another point of view.

Chicken, even in these days of her maturity and wisdom, is scared of everything. Spike, with the exception of sparrow-hawks, ladders and the Portsoy snake, was scared of nothing. No other object daunted him, no lawnmower, no hairdryer, no electric drill. Everything had to be approached, examined. In response, every movable object had to be stuck down or removed. Experience taught us what could safely be left on the table or worktops and what could not. The former category narrowed itself quickly to lump hammers, cast-iron pans and (had we had cause to place any there) objects made of concrete, marble or granite. The latter category included virtually everything else. Small though he was, he developed the strength needed to precipitate most items off the table or work surfaces to smash on the tile floor. He'd push cups to the edge, pause speculatively before giving a final nudge and then stand back waiting as the smash came, walking to the edge to peer over and look at the results. Small coffee cups, easy items to over-look, suffered the most. By the time we had accustomed ourselves to greater care, an entire set was destroyed. We learnt eventually not to leave anything – glasses, plates, cups – anywhere whence they might be launched to certain destruction. Anything requiring a permanent place, tea jar, candlesticks, everything, was stuck down with Blu-Tack. He would run across the table, over cutlery and plates, sample from glasses, perch on the edge of any left unguarded, dipping his beak in to drink. We had to be quick when alcohol was involved, his sober behaviour being sufficiently exuberant. He loved anything with bub-bles, mineral water or lemonade, reacting swiftly to the opening of a champagne bottle, carrying out a small toenail-tapping pavanne on the worktop with an eager celebratory air, appropriate for any occasion. (An old Tyrolean belief was that drinking the broth in which a magpie had been boiled would cause madness. I still wonder if, according to the principles of homoeopathy, inadvertently sharing a glass with a magpie might have a similar effect.) He stole and cached ice-cubes.

After we devised the system of establishing his bedtime, sending him to bed before we ate, as one might with a toddler, in order to have an evening of calm and peace, we began to test his skills of observation by refining the process, reducing what we said, from calling, 'Spikey, into bed!' until we needed to say nothing at all, until a glance would do; eventually it seemed that all it took was the thought for Spike, standing watching from above, to fly across the room to his box. He, of course, had observed the slight alteration in our manner of which we ourselves weren't even aware. He began to know by himself that on the second playing of *The Archers'* theme tune, it was time to launch himself bedwards.

His box was commodious, the size of a child's playpen, equally well stocked with what we had given him to play with and what he had found. We'd close the flaps loosely over the box after dark. He seemed to like both being enclosed and having the freedom to fly out at will in the morning. If the doors between rat room and kitchen and the rest of the house had been left open, any one of us might wake to see a small, curious black face, a pair of scrutinising dark eyes, peering round the bedroom door.

In time, too, we discovered Spike's delight in books. Until I learned better, I would leave the pile I was working on or reading on the kitchen table. His passion was, I suspect, more sensory than literary but it was love none the less. The physical qualities of books delighted him, the secret interstices that lurk between pages, the presentation of a manifold profusion of caching sites. The turning of each page was an opportunity, each a place where an item could be stored safely, unseen, remembered: half a prawn, a piece of Brie, a tiny sliver of cod. He loved the sensations of paper, the feel and sound of ripping, the unimaginable delight, after effort, of watching the scatter and fall from the table's edge of a fine shower of magpie confetti. He destroyed two pages of a library book on a very sombre topic I had inadvertently left on the table. (An act of financial contrition was required.) Konrad Lorenz talks of an animal's capacity to cause damage being proportional to its intelligence. How true.

We had decided in the usual arbitrary moment that Spike was male. It's difficult to differentiate by appearance between male and female magpies; females may be slightly smaller but to make an attempt at a correct assessment we'd have to have had a line-up of other magpies to compare. My use of 'he' should suggest no judgement on the basis of behaviour. I don't know if female magpies display any more discipline and sobriety than male ones. After I began to look closely at other magpies, I saw that Spike was quite small, but his unorthodox upbringing may have been the cause rather than his sex.

The feathers of Spike's tail, whilst always relatively short, became easily damaged. With difficulty and considerable danger to myself, I'd catch him and, clamping him under my arm, trim his tail feathers neatly with scissors, cutting off the bent and broken ends. When time and mince had dealt with his outrage, he'd celebrate his neat new tail with wild delight, chasing the tidy, shortened version in a whirling, prolonged and dizzying *kazatski*.

His vocabulary grew. His enunciation became more precise. Han would sit in the late afternoons at the table, studying for exams. Spike took a close interest, flying down onto her pages, trying to steal them or rip them up. Han would shout at him, bat him away but he was, as ever, indefatigable, returning time and again, beginning to lose his temper, muttering increasingly loudly as Han lost hers.

'Bugger off!' Han would yell.

'Bugger off!' Spike would shriek back, glissading in direct and awesome confrontation down the long wooden surface towards her papers.

(I have a sheaf of cuttings sent by friends with reports of foul-mouthed birds: the magpie who, after life with a north-east fisherman, had a wide and well-used vocabulary of vibrant oaths; the parrot who, in colourful, explicit terms, dealt neatly with the pillars of society by swearing in turn at mayor, policeman and vicar.)

The sounds Spike made, apart from obvious words, were both expressive and useful. 'Eh!' he'd say. 'Eh!'

'Eh!' Han and I would say to each other, and still do in greeting or agreement, a sort of verbal high-five, an expression of frustration or dubious acceptance. 'Eh!'

There was 'oy' too. 'OY!' he'd shout. His 'Oy!' covered a multiplicity of purposes, two letters containing a world of history, in both human and magpie terms, an accretion of aggrievedness, outrage, pathos.

One day, when he was a few months old, I came home to what I thought was an empty house to hear a conversation in progress in the kitchen. Standing outside the door, I listened. Although I have tried, there is no method by which I can render what I heard in words or letters. They were words but not words, a cascade, a trill, a babble of sounds; all the terms of the phonetic dictionary, rolled and lateral and fricative, syllable, consonant and vowel, were there. The voice was enthusiastic, eager in intonation, the rising, falling cadences of speech, the perfect pauses, an amazing mimetic model of speech. The words, for they were words of a sort, were a magpie's, but the voice was mine. It was the sound of me, conducting a phone call or conversation, every nuance of intonation, every laugh, every mannerism, me, revealed through the conduit of a magpie's voice. Spike spoke in my voice and, more than that, he laughed with my laugh. I listen to the tape I have of him and hear myself, piped through the vocal cords of a bird.

'Hello, Spikey!' the voice yells, with the faintest suggestion of the interrogative in the second word. The sound is ethereal now, otherworldly, almost as if I no longer know whose voice it is. Recently I played it to friends.

'My God,' they said, awed, impressed. They looked at one another, wondering if they were about to say something I had never appreciated. I saw them stifle the urge to laugh. 'He sounds like you,' they said, as if I might not know.

Pliny suggests that for magpies speech is an obsession, that death results from their failure to enunciate a chosen word, or that they will cheer up after hearing the sound of a forgotten word. Having known Spike, I suspect it may be true.

Whilst most of Spike's speech appeared to be mimetic, the one word he consistently used correctly (possibly also in imitation) was 'what'.

I would go into the kitchen and when I didn't see Spike, call him, knowing he'd be in one of his hiding places on top of the kitchen cupboards. 'Spike?' I'd say.

'What?' a voice would shout from above. 'WHAT?'

It's reasonable that, were he going to speak like anyone, it would be me. We spent a lot of time together. I talk to all the birds and nothing escaped Spike's attention, neither sound nor action. I noticed that Chicken too was affected as the impression of words became noticeable in her voice, the faint imprint of 'hello' amid her usual calls as I believe she tried to emulate him.

Certain domestic tasks intrigued Spike, the making of bread, the cleaning of any fish, brass- and silver-polishing, when he'd fly down from the top cupboard, shouting greetings and general enthusiasm, to stand beside me on the worktop to watch and contribute his inimitable help, my unhelpful helper. He'd knock over the smaller items I'd polished, then perch himself on the edge of my empty coffee mug to watch the progress of the matter, trying all the while to steal my cloth, hopping with annoyance when I resisted, snapping at me. If I was making bread, he'd make holes in the flour bag. He'd dance about trying to get his beak in to drink the egg I'd beaten to glaze the loaves. When flour, yeast and water had been kneaded into dough, he'd beg and squeak, bearing off his portion to secrete in a place where later, if I got to it before he did, I'd find it, risen and squashed behind my oven glove, in the folds of a teatowel, behind a cushion on one of the Orkney chairs.

Spike had the advantage over Chicken in his caching activities, since Chicken has never had the opportunities afforded by flight. Spike was able to cache in a multiplicity of sites, whilst Chicken did as she has always done, hid her food under the carpet, whence it was routinely retrieved by Spike, or in her hole in the wall. Chicken regarded the theft with resignation, waiting until Spike was in bed before doing a round of his accessible caching sites to redeem her stolen items. Often, I saw her looking in unexpected places, behind the coal scuttle or under the corner of the fridge, places where I knew she didn't cache. As all the research I've read subsequently suggests, where caching is concerned it takes a thief to recognise the malign intentions of another one. It seemed that, when it came to honesty, there was nothing to choose between rook and magpie.

For them, caching wasn't a matter of necessity, since they had food available to them at most times (Chicken particularly, because she had her own quarters in my study where Spike was not allowed to go, for the sake of my computer, my papers, the furniture, the lamps, the curtains, as well as in respect for Chicken's privacy). Spike appeared to cache from a spirit of general busyness, because it was, and is, what corvids do. Spike too, when the chance arose, appeared to do a routine survey of both his and Chicken's cache sites, checking, removing, replacing. I'd watch his methodical progress. His memory was impressive.

Chicken was never in any doubt about Spike's potential behaviour. During his infancy I kept them apart because, while we were enthusiastic about fledgling magpies, it might have been expecting too much to assume that Chicken would share our feelings. I could imagine all too well the damage an adult rook's beak could do to an unfeathered infant. When Spike grew to the size where he was more than able to defend himself, I never left them alone together because by then it was Spike I didn't trust. Apart from his possible intentions, he was too energetic, too lively for the increasingly sedate Chicken, whom he'd chase round the kitchen table with brisk determination, nipping at her tail. Chicken, a sociable sort of bird, showed only the mildest annoyance at his behaviour, snapping her beak towards him in token protest. They were mutually wary but would on occasion co-operate, sitting together on someone's knee or the same chair, Chicken occupying the seat, Spike the back.

Konrad Lorenz describes a magpie of his acquaintance as 'a feathered rascal, lacking any sense of propriety'. Where Spike was concerned this was indubitably the case, although occasional evidence showed at least uncharacteristic moments of an intermittent propriety.

Spike played prodigiously. We gave him toys, a small doll which he dragged about by its bright yellow hair (a good luck troll, donated by Han. I have it still, rendered bald by magpie attentions), a ball which he pushed with his beak and chased. He appropriated from somewhere the empty spherical container from an air freshener, to which he became unfeasibly attached. He would fly the length of the kitchen carrying something demonstrably too heavy for him in his beak, a wooden spoon, a ruler. He loved to fly, carrying the paper butterfly I had bought for him. He found a tiny blue plastic shoe, a fairy doll with gauzy wings. He'd steal chopsticks, carefully measuring them with his eye before picking them up, perfectly centred, to fly with them across the room. He loved ball games and would play by himself or with whoever would play with him, on the kitchen floor. He and Han would run the length of kitchen to rat room and back again, exchanging the ball, Spike squeaking his enthusiasm, flapping his wings, shouting with what seemed remarkably like joy.

Han, at the time a fervent practitioner in the martial arts, spent time practising kung fu, travelling to competitions from which she returned with ever larger and more fantastical trophies. (The one which accompanied her from Florida when she was about fifteen, her seat companion on a long, cramped flight, was taller than she was. An elaborate, tottering structure of cheap gilt, pilastered and curlicued, topped by an ornament alarmingly like a golden funerary urn, stands still, slightly unsteadily, under a light coating of dust in her room.) It became her habit of an evening, before Spike's bedtime, to engage with him in a bout of combat, an enterprise that delighted him since he was unfailingly up for a fight. She would initiate the bout by punching the air near his head, one side, then the other, just enough to enrage him, enough to cause him to fly to the top of the fridge, where he'd stand quivering, readying himself, waiting for the moment when he'd attack,

it was intellect that glittered from his eyes

his the advantage in proper flight, hurling himself towards her, eyes yellow and protected, squeaking with martial fury, wings a blur and rustle of crisp, bright feather. Wham! Wham! He'd squeak frenetically, shouting random words – 'Smike! Oy! Oy! Spikey! Hello! Hello!' – as he attacked her moving fists, diving for her head as she leapt and danced away from him. They spun and fought, Spike launching himself again and again towards her, at her, a marvellous, wild, balletic frenzy of black and white, all the more strange and thrilling perhaps because of the imbalance in size of the participants, their cultural diversity, or the fact that one of them at least had failed to master the important philosophical requirements of the martial arts.

Spike's beak, adapted for the multifarious purposes of magpies, could be alarming. 'He'll take your eye out,' both girls said to me often. I never believed he would but, unusually charitably, thought that if he had it wouldn't have been on purpose. He liked to sit on the high back of the dining chair to allopreen my hair. I was flattered but concerned. The force of his beak was much more powerful than Chicken's and the memory never far of a revolting scene from Circus Lumière that I saw many years ago at the Edinburgh Festival, which involved a man miming the consumption of his own brain by way of a spoon dipped into the top hat he was wearing, collapsing further with every spoonful into a state of alarming, slobbering, decerebrate unconsciousness.

Not the picky gourmet that Chicken is, raw squid was Spike's particular treat, the soggy fishmonger's bag a sight to make him mutter and tremble with covetous delight as he waited for the moment of revelation, the sight of the slippery white body, purple dotted tentacles, slimy pouches of entrail. He'd stand and flap his wings in barely controlled anticipatory ecstasy, eyes rolling in alarming yellow circles, drooling horribly, saliva dripping down the feathery fringes at the sides of his beak, as I tugged off the heads with their beseeching, nacreous eyes, and emptied the envelopes of tough white flesh of their sandy, muddy contents, disclosing the tiny fish within which he'd lunge for and gobble. I fed him discarded pieces and he'd fly up to the top of the cupboard, trailing things disgusting and slippery from his beak. He cached them resolutely, committing to memory each ghastly morsel, some of which I'd find later, warm and beginning, ever so gently, to putrefy.

In common with many birds – owls, crows, ravens, raptors (but not Chicken, I'm not sure why) – Spike disgorged pellets from his throat and beak. The disgorgement was the end-game of a ritualistic and (to the hapless observer) disgusting spectacle of eye-rolling, drooling, gagging, a process that reduced Han to wails of horror. The pellets appeared to be seed cases, grit, bits and pieces of unidentifiable detritus, glued together into a compact slimy bullet.

One of Spike's pastimes was to fly to the spare room, where he could stand on the edge of the Victorian cot and gaze out over the garden.

'Do you want to go upstairs, Spike?' I'd say, and he'd fly towards the door, which I'd open for him, round the corner, up the stairs and turn into the spare bedroom, where he'd take his place on the brass cot rail. One day I walked into the room and saw Spike standing, as he liked to do, gazing from the window into the eyes of another magpie who stood on the outside sill, gazing back. He was not afraid. Nor was his compadre, or comadre, on the other side of the glass. (Spike seemed to understand glass, I think, a little more than most birds but still screamed in fear at the sight of a sparrow-hawk in the distant sky.)

In the same way that he would fly upstairs when told to, he would – when he chose – come when he was called, materialising in an instant on the worktop beside me. When he didn't choose, he had to be pursued, then found, not an easy matter because a magpie intent on hiding can form itself into an imperceptibly tiny shard of magpie and insert itself into the narrowest space behind beds, in corners, under chairs. The moment I found him, he'd escape, run and launch himself gracefully over the banisters.

Like Chicken, Spike too bore out Konrad Lorenz's contention that corvids hold a deep, lasting suspicion that many things black are the corpses of unfortunate corvids. Spike, though, was more outraged than afraid. Regularly, he'd attack a pair of scruffy black and white trainers I used to wear, on the grounds, presumably, that he suspected me of a crime of which I have never been guilty, that of wearing on my feet a pair of flat magpies.

Upstairs I have a stuffed-toy rook, as perfect in every detail as a stuffed, fuzzy black rook can be, from its stance to its eyes, its grey felt beak to its faintly curled toes. Made by a German toy company, it was bought for me as a birthday present by Bec, in a toyshop in Amsterdam. When I brought it home, I showed it briefly to Chicken, who was scared and displeased. I don't know if she observed it merely as something black and unwelcome, like Konrad Lorenz's bathing drawers, or as another rook, a rather static, stately one, but a rook none the less. Whatever she saw, she shouted vigorously and ran from it, clearly frightened and perhaps insulted, and I have never shown it to her again. I can only guess at what her reaction indicated: that she considers fuzzy German replications of rooks possible rivals, or that, for her own reasons, she'll have no truck with graven images. She is however unconcerned by seeing her image reflected in glass, in the surfaces of kitchen equipment. She gives no indication of thinking, as many other birds would, that the reflection is another rook.

Spike too seemed interested by catching sight of his own reflection in a window or mirror, reacting with studied attention, as if he was appraising his own appearance, a reaction that bears out the findings of the research into magpies and mirrors quoted in *In the Company of Crows and Ravens*. Even then, before I knew of the research, I'd wonder what it was that Spike was observing and what it might mean, or indicate about the nature of his consciousness of self. That he had a high degree of consciousness of self was something I never doubted.

Consciousness in humans is a concept that defies definition, evades attempts to explain precisely what it is and how it may be described or manifested (rather like intelligence, but even more elusive). There are degrees of consciousness, from primary consciousness, an awareness of our own being, to higher-level consciousness, what Gerald Edelman describes as 'the state of being conscious of being conscious'. The neurological basis of human consciousness is being constantly analysed, examined, taken apart, the neural components and structures of what makes us thinking, imaginative, creative, self-examining beings, a dissection of the ways in which the material workings of the brain give rise to the immateriality of thought. Complex to describe and interpret in humans, how much more so it is in animals or birds. The evolutionary advantages of consciousness are clear, allowing anticipation of the behaviour of others, adaptations in situations of social complexity and change. The area of the human brain most intimately associated with what is broadly described as the mind is the neocortex, a structure similar in function to the avian nidopallium. Since the evidence appears to demonstrate that the anatomy of the brains of birds is sufficiently similar to that of humans to allow both cognition and consciousness, it doesn't seem unreasonable to believe that some birds have a degree of consciousness of self – although since the purely physical examination of the mechanisms of the brain doesn't fully explain consciousness in humans, it's not likely to in birds either.

Because, where corvids are concerned, I am prepared to consider anything, I did think, from time to time, that Spike was cleverer than anyone or anything else I had ever met, that it was intellect that glittered from his eyes, radiated from him, a marvel trapped within his tiny feathered body. More than just conscious of himself, he seemed also to have a distinctive sense of others, bird and human. It was as if Spike had an eternal patience, a self-knowledge, a confidence that told me that he, at least, believed that his and his kind's day would come, that when we, humankind, had finally wrought our considerable worst, it would be then his turn, the turn of birds.

Louis J. Halle describes holding a swift in his hand and noticing the blankness of its eyes. It is, he says, impossible to imagine communicating with one as with a parrot, duck or canary. 'I suspect that the swift inhabits a world of birds and insects, in which people are marginal . . .' (Swifts live on the wing, copulate, feed and drink in flight. Our world and theirs do not meet.) Spike had beautiful eyes (which may or may not have signalled intelligence and depth) and enviable film-star eyelashes. On the frequent occasions we were close, Spike on my knee or standing on the pages of the book I was reading, we'd look into one another's eyes in silent, inexplicably profound conversation. He would hold my gaze, look at me straight for a long time, and when he did I knew him in every respect my equal, more than my equal: he made me aware that my only advantage was one gained through evolutionary time.

Often I'd sit beside the stove in the kitchen, Chicken on my foot, Spike on my knee or shoulder, and do what I still do with Chicken, ponder our courses in life, seeing the division between us as nugatory, an illusion, the three of us forming an unlikely, harmonious triangle. There could be a silence between us, an ease that once I would never have thought possible, had I even been able to imagine it. To believe that humans have a monopoly of the things that deepen life on this earth – memory, appreciation, imagination, emotion – seems both arrogant and simplistic; to imagine that, without a language we recognise, birds and animals exist in a world of thoughtlessness, of lesser communication, lesser feeling, surely wrong.

In time, Han left home. Spike, at my behest, would phone her at her university flat in Edinburgh. 'HELLO!' he'd yell down the phone as I held him in my hand, 'HELLO!' and she'd call her unbelieving friends to listen to this wonder, a talking, human-voiced magpie.

One day, when I was carrying a ladder through the kitchen, Spike, in fear and panic, flew across the room and collided with the wall. He fell to the floor, stunned. I picked him up and put him into his box. For days, he was huddled and listless. I fretted over him and, when he regained his interest in food, hand-fed him. Through good fortune or the sturdy qualities of the magpie constitution, he recovered. Birds die when they fly into windows or walls, not, as commonly thought, by breaking their necks, but by damaging their brains. Spike's brain seemed, after the incident, as vibrant as it had always been.

A few years after Spike came here, I read in a copy of a wildlife magazine an article by Professor Tim Birkhead, author of a definitive book about magpies. He suggested that magpies, being 'too obstreperous for domestication', cannot successfully be tamed, that, not being subject to the long processes of domestication undergone by cats and dogs, like all wild creatures their own natural inclinations will inevitably conflict with what he describes as 'civilised restraints'. As I read it, I felt uneasy, sensing in it something I already knew.

Indeed, after a few years Spike grew in criminality. His aggression expanded, his area of operations, his private gangland territory, to encompass the cupboard where the dustbin bags were kept, the door of the dishwasher, the valve of the floor steamer. He flew with greater, more malicious intent. I could see the reluctance of some friends to come into the kitchen. 'Just shoo him away,' I'd say, but would note their wary, sensible fear. I had read Professor Birkhead's article. I knew him to be right.

If people are frightened of Chicken, they were even more so of Spike. I was careful to keep him away from each nervous, cowering joiner, each terrified electrician, each visitor, imagining, as I chased him from kitchen to rat room, slamming the door between them, the lawyer's letters arriving, accusing me of failing to keep proper control of a magpie.

I worried about the future. Unlike Chicken, he showed no sign of maturing into a contented adult bird. I knew him to be too vigorous, enquiring, demanding for the life he was being obliged to lead. It felt as if he had other things to do, other places to go. It increased my sense of guilt. What would have happened to him if Elizabeth hadn't phoned me that day? For most of the time he was here, I believe that he was happy. He was, as far as I could judge by his behaviour, his appearance, his interest in himself and in what was around him; but it was not enough. As he grew older, he wanted to be elsewhere. I had the same misgivings and anxieties as I did with Chicken. I might have decided differently about him had I known, but without him we, at least, would have missed so much.

It was only months after I began to wonder what would be when I came back to find Spike in his box, crouching, clearly unwell. He had been in fine and vigorous form when I had left. Within a day, he was dead. I don't know why he died, what it was that made him wilt, as birds, do, decline within that very short time and die. He was as he had been when he had flown into the wall in fear at the sight of the ladder. I can think of no other explanation. Frightened by something in my absence, a cat at the window, the patrolling hawk, he may have flown against the wall. He drooped, beyond my ability to help, his eyes clouded. I sat on the floor by his box, stroking him.

I wept the night he died. Sitting in bed, filled with the utter loss of his person, I felt diminished, bereft. I talked about him, but not very much, in the main to members of the family, who felt the same, but to few others.

It's the only way, this compact and measured grief, for those of us who are aware that there has to be proportion in loss and mourning; we laugh at ourselves for our grief, trying to deal with this feeling that is different in quality, incomparable with the loss of a human being. I laughed at myself as I had once laughed at and with a friend, a surgeon, while he told of weeping in the operating theatre the morning after the death of his parrot. I laughed but I cried too, briefly, in the telling of Spike's death, in the staffroom at work. I cried, mourned him, as we all did. We felt – we knew – that something immeasurable had gone. Han felt it particularly, his sibling and his friend. As it may be after any death, of someone, something close, we may lose a part of ourselves, however small a part, one that anchors or measures us, places us in our particular constellation, if not among the stars, then in our reflection on this earth.

Han and I, in particular, both still miss him. We talk about him often. 'Eh!' we say to one another, 'eh!'

I recalled again this sense of closeness one night recently when, just towards midnight, I heard Chicken call. I ran downstairs to where she was standing on the floor, distressed, appearing disorientated. I didn't know what had happened. She seemed to find it difficult to hop back onto her branch. I lifted her onto it and for the rest of the night lay between sleep and waking, afraid of what might be, and during the hours of the night I knew again that we are by now part of one another in a way I cannot easily explain, that she defines something of my life, as Spike too had done. I came down early, six-ish, filled with anxiety, and as I reached the bottom of the stairs heard her, in full vigour of her morning voice. Perhaps the dove who insists on sleeping on the ledge of the window above the study had disturbed her, or the light of the full moon had startled her from sleep.

I think about Spike often. Walking through the district, I see magpies, hear their distinct, distinctive *chuk-chuk-chuk*. I think of the vibrancy of his, of all magpies' existence. I think of what he was.

A couple of years ago, a friend gave me a book of poetry. In it I found the finest, simplest counterweight to all anti-magpie sentiment. In 1917, the year before his death at Hermies in the Somme valley, a Derbyshire vicar's son, Theodore Percival Cameron Wilson, wrote:

The magpies in Picardy
Are more than I can tell
They flicker down the dusty roads
And cast a magpie spell
On the men who march through Picardy
Through Picardy to Hell.
(The blackbird flies with panic,
The swallows go like light,
The finches move like ladies,
The owl floats by at night,
But the great and flashing magpie,
He flies as lovers might.)

Part III

BIRD SEASONS

rooks fly overhead, burdened with twigs

12

Winter

On a morning in October, as I'm opening the doo'cot door early, I hear the geese returning. It's the first sound of winter. If there has been a summer, it's over by September, the few bright days full of intimations of cold, the nights beginning to frost. The scent of wood- and peat-smoke is blown on the wind. We are far north here, north of Moscow, Riga, Novosibirsk. Almost surrounded by sea, the intensity of cold is less than on the great land masses of northern Europe and Russia, but the seasonal light is the same: long, glowing days and nights followed by diminuendo into darkness.

In the newly chill mornings, the geese are streaming back from their summer homes in Greenland, Iceland, calling, trailing black, drypoint lines across the high northern sky. They are pink-footed and barnacle geese flying in the formation that gives safety and guidance to the young, the new migrators, the paths of all but the leading bird eased as they gain lift and are pulled into the circular air currents spi-ralling from the tip-vortex of the wings of the bird in front. I stand and watch them pass overhead, listen to their voices. They feel like part of this place, part of the season and the sky. They come back to the flat land, to fields and water, to the Loch of Skene a few miles away, the Loch of Strathbeg forty miles north. For some, they're a bane, symbol of the many darknesses of winter, but I welcome them. I like to know that there's something, someone who chooses to be here, to share our winter.

Watching the geese, I think of migration, of the leavings and arrivals, of the many mysterious ways in which migrating birds know where to fly, of the dangers of their journeys, these intrepid, brilliant international travellers. Migration is unimaginable, the distances, heights and speeds. (The swallows and swifts have long gone to their winter grounds in Africa. Their going is the beginning of the winter's silence.) We are all, bird and human, part of the earth, of its time and its matter, impelled by the mechanisms within, the ones that order our responses to days, months, years, to light and darkness, the rhythms, circadian, circannual, that regulate what we are and what we do.

It's the genetically determined activity known as *Zugunruhe*, migra-tory restlessness, that controls migratory behaviour in birds, making even caged birds turn, fluttering, dancing, towards the direction in which they know they should be going. Birds migrate in response to the urgencies, the needs of their lives, for food, or warmth, or for the con-ditions that may help to ensure their continued survival. They may make limited migrations, moving vertically from lower to higher ground according to the season; they may cross continents or span the world. The patterns of migration may be variable, some birds migrating only in response to a seasonal lack of the specialised food they require.

Preparatory to migration, a bird's metabolism changes, its response to day and night. To develop sufficient fat to allow their often pro-longed flights, they eat much more than usual, more fruit, more high-calorie food. Their bodies become more efficient at using food and they accumulate quantities of subcutaneous fat, greatly increasing their percentage of body fat, their fuel for the daunting, hazardous journeys they undertake. The cost in mortality of migration is high. Migrants face inestimable dangers – exhaustion, storms, becoming lost, landing in the wrong places, interception by predators, by hunters, starvation, drought – but still they do it, obliged by the clock within.

Most birds migrate at night. The ones who don't are obliged by the nature of their wings to take advantage of warm air to aid their flight. The night flyers use the calm and cool of night air, leaving shortly after dusk, coming to land before daybreak. They fly in great flocks, or they fly alone, a single bird, a family group.

Most birds fly at low altitudes, between two and six thousand feet, but many, particularly waders, fly higher, at fifteen to twenty thousand feet, while the mountain birds, lammergeyers and bar-headed geese may fly at altitudes of twenty-five to twenty-eight thousand feet.

The long-distance star of migration, the Arctic tern, a smallish bird, breeds in the Arctic and migrates eleven thousand miles to the Antarctic. In a year, this small bird traverses the world.

Rooks too migrate, though not in great numbers, mainly from northern European countries to Britain for the winter. At migration times, I watch Chicken for signs of *Zugunruhe*. I try to identify in her that urgent, ontological restlessness but I see no hopping, no straining to be off. She seems as peaceable as ever, contented, displaying no indication of being overly attracted by doors or windows, no particularly noticeable interest in either north or south.

In late October, the clocks change. (For those us of wintry disposition, whose circadian rhythms respond to this particular seasonal adjustment, the change is welcome.) In the bare garden, the doves peck among the empty stalks and bathe in the rain of cold afternoons. By now, they head homewards from the surrounding roofs by mid-afternoon.

It's Hallowe'en and then Bonfire Night. In preparation, I shut the doves in early. To muffle the noise a little, I close the wooden door over the screen of wire that lets in light. I speak reassuringly to them as they move uneasily on their perches within the deafening range of the fireworks' artillery thuds. They don't cower silently in fear. They sit along their perches, angry, expressing to one another loudly their dislike, their fury at the pointless noise, at the interruption of their peaceful nights. (They must be used to city noise, although I wonder what they think of the weekend routine of late-night Scottish cities, a never-ending scale, a continuum between rejoicing and lament as drunks band together against the world in song and chanting through the darkened streets. Whether the clamour wakens sleeping doves at two, three, four in the morning, those uncertain sounds, if they regard them merely as another of the vicissitudes, the frustrations, the end-lessly inexplicable conundrums of life in close contact with people, I do not know.)

One blowy morning in December, I find a stranger in the doo'cot, a sleek, beautiful, bewildered bird, one elegant pink leg ornamented by the enigmatic tag engraved with the indecipherable code of numbers that denotes a young racing pigeon. It has happened a few times over the years: a young bird, newly released to find its way, gets lost, deceived by magnetic storms or cloud and, I assume, on seeing my birds flying, follows them home. This one, like the others that have appeared this way, is a rare and cultured beauty, with the equivalent of the high-bridged nose of an aristocratic pharaonic lady. Clearly hungry, it feeds nervously. I shut the door while I catch it. I always regard these birds with a kind of fearful respect (they are highly bred and usually expensive), removing them quickly from the low company of my own rabble and putting them into boxes in the toolshed, scared that they might be impregnated by one of my common types, deter-mined upon class warfare. I examine the leg-ring and phone the appropriate organisation. Immediately, they identify the bird from the numbers.

Before a morose man arrives and silently, unsmilingly, reclaims what is rightfully his, I carry the bird to the kitchen in a box. Over the years, I've become used to visiting pigeon fanciers, the ones who work as window cleaners or joiners but are, at their root and in their bones, pigeon men. (There are pigeon women too, but men predominate in their passion and commitment. The disparity of sexes is accounted for, I assume, by the fact that in the main women have other houses to clean, other food to prepare, other offspring to whom maximum attention must be given.) I used to dread the way, whenever one was in the house, they'd stop at the first faint hint of a *roucoule* or a *kurre kurre* from outside the window, listen, run outside to peer at the doves, to examine their house, their food, their water, to give their opinions of what I was doing wrong, some of them managing graciously enough to conceal their contempt for the lax standards of my dove-keeping. Now I accept the distance between their expectations and mine, for these are experts, urban Mendels, easy with Xs and Ys, with dominant and recessive, with straight wings and stepped wings, meiosis, intragenic complementation, all things about which I know absolutely nothing. (I think of them too when the sparrow-hawk visits. I'm sorry when it's one of my birds but how might I feel if I had paid tens, if not hundreds or thousands, of pounds, watching it being silently consumed in the middle of the flowerbed, the worth of the remains, a pair of feet, a loop of coral entrails, a slew of greyish, dampened feathers?)

In early evening, the owner arrives and I hand over the bird. I don't like to think about it getting lost, being blown adrift by gales, led astray by the distortions of solar energy, landing at best in an alien doo'cot filled with potential maniacs. As I always do, I spent a little time stroking and admiring it, talking to it, wondering how much it might have cost. As with the others, I considered ransom.

streaming back from their summer homes

There are freezing nights in December and January. I worry about my very old doves, the few who must be twelve or fourteen, perhaps older, possibly failing of sight, tatty of feather or frankly balding, irascible, treated with wary caution by their younger housemates; but so far no one has perished from the cold. Few seem to die in their beds, or rather on their perches. They're not there one day, slower perhaps than the others, grabbed on the wing by the watching raptor.

I shut them in, taking my torch, or when Leah, Bec's daughter, is here she carries it for me, my intrepid torchbearer. We make our way to the doo'cot, shine the light in to look at the lines of roosting doves. They hold themselves very still in the cold air, fluffing their feathers, drawing themselves in, standing on one leg to conserve their warmth. (I remember the cold owl of Keats's poem 'The Eve of St. Agnes'.) In the morning, we'll smash the ice on the water dishes and bath. I listen in the darkness for the hum and *chwee* of the small electric fence a friend installed around the doo'cot roof for me to keep out intruders. I check it regularly. I like the sound of it as it sends out its little warning signals.

When it snows, I confine the doves in case they can't find their way home. They mutter between themselves but I don't know what their opinion is of snow. On these mornings I open the kitchen curtains for Bardie, whose house is placed carefully so that he has a good view of the garden. He looks out, and at the sight of the altered light, the white ivy outside the window, he shrieks like a delighted child with many loud, repeated exclamations of excitement. I like to think that he's expressing true Antipodean enthusiasm, a loud and hearty expletive at the wonder of it all.

This year there has been only intermittent snow. In the main it has been damp and dark and wind-blown. Chicken and I lurk together in the study or beside the stove in the kitchen. She climbs onto the arm of the chair beside my desk on the dark afternoons and stands close. In the evening, she warms herself under the reading lamp as she sits on my knee. She in turn warms me with her small, hot feet. I envy her her feathers. The wind screams around the granite of the walls.

In late winter, my cousin Roger visits. It's a long time since we saw one another last but we have things to talk about. That time was, by chance, 11 September 2001, also the first time we had been together for many years. We remember, retracing as everyone may do the circumstances, where we were, how we heard, all in the kind of close detail we achieve over nothing else. On this occasion, however, we want to go to look at birds.

We drive in cold sunlight up the bleak north-east coast, through the dissonance of a countryside half agricultural, half industrial, the flatlands between Aberdeen and Peterhead and Fraserburgh, all fields and oil-related sites, factories and yards stacked with giant lines of pipe, equipment for locating and drilling and making the undersea yield up its oil. We stop at the harbour in Peterhead and park on the slipway to watch a ship prepare for sailing. Everywhere, gulls scatter and scream in the chill wind. Our aim is to visit geese.

We turn down the road towards the Loch of Strathbeg and park under the awning with its wooden stanchions. We are the only people here. It has felt intermittently today as if we're the only people anywhere. There is a farm building, now converted to a hide. We look out of the broad windows over the loch. The literature on the table tells us the facts, the numbers, that 20 per cent of the world's population of pink-footed geese are to be found here at this time of year. On this particular day, though, they're not. I don't know where they are, the 20 per cent (which is many, many geese. On some days, thirty thousand.) I don't know where they've gone but they're not here. The loch is empty. Like us, they have gone out for the day. No one, no bird, stirs. There are no whooper swans, no wigeon, no teal. As bird watchers, we have failed. The grey water and reeds are stirred by wind. There are no geese.

We make coffee from the machine and, clutching the plastic cups to keep us warm, sit and peer through the fixed binoculars, although we have our own. Beyond us, the timeless calm of winter. Just past Imbolc, the Celtic winter festival, still in the wolf-month, that reminder of the terrors of the past. The window seems a symbol, invisible glass between us and the water and the winter sky. There are no birds and then there is one bird, a black and white duck which appears on the surface of the water. We decide that it's a tufted duck although it's quite possible that it's not. We discuss what it might be and come to a far from certain conclusion. From the grey attenuated light of this February afternoon, a heron assumes bird shape from a haze of flying shadow, drifts down to stand with ancient, ancestral dignity, his air of grey, ecclesiastical solemnity, in the shallows in the fringes of reeds. We watch him intently in silence for a long time, until at last he takes off with slow, considered grace, flies low to merge languidly back into the faint winter mist from which he emerged. Ahead of us, empty water and quiet. We are satisfied. There is no disappointment as there was no expectation. This is the obverse of bird-watching. We are in a place beloved by birds, chosen, for this time, by us. Birds just happen, today, this afternoon, not to be here.

We carry on. Farming country, the quiet time. Cliffs fall, sheared from green fields, the roads long and quiet unlike most other roads, ones further south, choked and clogged with traffic. Pensive hawks stand on fence posts, circle in their slow watching above us. We drive to Pennan, down the terrifyingly steep, curving approaches, looking down, like birds ourselves, on to roofs and chimneys. It's the village made famous by the film *Local Hero*, quiet and perfect; and, just as there were no birds at the loch, there are no people in Pennan. The village round its tiny bay is closed for winter, silent except for the sounds of sea on stone. The sun's bright although it's freezing. Ice travels on the wind. The famous phone box is there but no one is phoning from it, as people do, to Japan and America, to say that they're in that phone box, the one in the film. Roger, a film director, explains how the filming would have been accomplished. Two large crows fly round the headland towards the cliffs. It will all awaken again in spring but just now it's like a Sleeping Beauty village and I wonder if everyone's really still there, inside their tiny fisherman's houses, asleep, suspended, liminal, in diapause, between worlds, between seasons, behind their small windows and their whitewash, evading the winter storms, the power and anger of the waves only a few feet from their walls. It is a village under a spell, under an enchantment of sea and sky and winter.

We drive back to Aberdeen in early dusk. (The geese will be returning to the loch. Our timing was wrong. We were too early, too late. They will be flocking in their hundreds, their thousands, calling and crying, their wings sounding like beaten metal.) Roger peers at his sat-nav while I tell him the way. We don't agree on every point, the sat-nav and I, but I bow to its wisdom and between the two of us we get here, home, after a day I will cherish, of cold, of winter sun, of laughing at everything, the occasion when together Roger and I failed to see 20 per cent of the world's population of pink-footed geese.

13

Spring in the House of Rooks

Late February. It is spring. Few external signs, neither the temperature nor the unrelenting grey of the sky, even hint at it, but it is so. Authority for the assertion is impeccable, beyond question. True, it's a little warmer than is usual at this time of year, although from this I draw no conclusions about wider climatic upheavals. Last year at exactly this time, there was so much snow that I took a photo from the study window on my phone and sent it to a friend, accompanied by a cryptic message which might or might not have suggested that I was in St Petersburg. He didn't recognise my garden under the unprecedented depth of late winter snow. 'When do you get back?' he texted in reply, but this year we have had, apart from a derisory inch or so lasting for no more than two days, no snow. It rains instead, a grey, intense, wet rain, attended by a fierce, bone-gnawing wind.

I know that it's spring because Chicken alerted me to it. It's three weeks since I recognised the first intimations, the sound of plastic against wood and tile as she began the inexplicable, now annual ritual of removing her food dishes from her house to clatter them across the study, the hall, the kitchen floors. The sound began early one morning and has continued, with interruptions for shouting, eating, sleeping and nesting under the dining table, more or less ever since. Most of the rest I understand, but the role and significance of food-dish-clattering as an indication and necessary activity of spring remain obscure. The clattering stops only when she takes up her other, alternative spring obsession, that of tugging at the wires of the ineffective anti-cat sonic device that, years ago, I rigged up on the outside lintel of the study door, long disconnected because of its general uselessness in scaring cats, left there only because it has the sinister dark air of a powerful yet unobtrusive security device. The disconnected wires enter the house under the garden door of the study, lying hidden behind the curtain, until spring when they are tugged, dragged across the floor, tugged again, shaken, dragged back, then tugged again. Chicken is about to begin nesting.

I don't know how Chicken knows it's spring. She was a fledgling when she first came to this house, still in her own first spring. I don't know what she thinks, how she discerns, what she remembers. I don't know how spring stirs in her. I have no doubt that she knows, senses, as we all may do, what she has missed or never knew, and for me to suppose otherwise would be to regard her with less respect than I do myself. This spring, as every spring, when it becomes clear that her yearnings are for other things, for mating and reproduction, I'm sad, guilty, regretful. I have obliged her to trade nature for her additional years. I can't, at this time of year, be sure that she considers it a suitable exchange.

Chicken's response to the first, invisible, and – to me at least – undetectable stirrings of spring leaves me bewildered every year. She lives indoors. What is visible of the world is through the windows – from her viewpoint, grey sky and rain, the tangled, trailing stems of a leafless clematis, the empty branches of the next-door beech tree – and yet she knows. Some clock within awakens, then wakens her. The sound of the outside birds perhaps, or more likely the increasing length of the days. The temperature of her existence is dictated by me, as are the switching on and off of lamps, the times of lightening and darkening of rooms, but since she will not allow the closing of curtains she's aware of the first beginnings of morning, the fall of night. Outside, it is still winter. The clocks have not yet changed. How does she know? The moon, new or full, waxing gibbous, waning gibbous, glows in its nightly arc, progresses in its pearled diminuendo beyond the study windows, through the branches of the apple and the pear trees against the wall before it slides off into invisibility above Queen's Road. Is it the moon that tells her? Is it a sense of the day, or night? Does the vernal equinox, just past, send her invisible, spectral, lunar messages? Does the slow, precise alteration in the slant of the rays of the sun communicate with her in ways I cannot even imagine?

After the initial dish-clattering, she begins to show other signs of what is clearly preoccupying her. Spring has always provoked sexual display as she bends, fans and quivers her tail, spreads wide her wings, lifting them, dropping the edges low, behaviour that is common to both males and females. Her behaviour is not specific; she does not display only to me, the person with whom she spends the most time. Anyone will provoke the same response, male or female. She displays particularly when addressed. She excretes copiously (even more than usual), on the rugs, the wooden floors of hall and study, the tiles of the kitchen.

Alerted by Chicken, I begin to pay attention to the garden. I notice, among the small birds, the robins and sparrows and blue-tits, activity of a kind that indicates that they too have noticed something in the season that I haven't. A lubricious play-chasing is taking place through and round the viburnum, over the dense branches of potentilla. Voices are heard in loud, calling excitement above the sound of rain. I have noticed sparrows carrying twigs. There are no leaves yet on the *Hydrangea petiolaris* on the wall to conceal the old, abandoned black-bird's nest which is exposed and empty, a small, useful store from which birds come to remove sticks and straw for use in their own nests.

There are the beginnings of coquettishness and display, even in cold rain, on the roof of the rat room as the doves limber up for spring. Males huff and puff and march, with an air of solid certainty of their own irresistible beauty, across the wet and gleaming slates. For me this must be the beginning of a time of vigilance if I don't want uncontrolled population explosion.

Last summer, for the first time, a pair of doves moved from the comparative comfort of the doo'cot to nest in the thick ivy which we've encouraged to grow on the house wall to ameliorate the unrelieved greyness of the granite. Over the years, it has grown up over the wall behind the kitchen, outside the study, become a rich habitat, a kind of tenement for birds, equal in its well-ordered, vertical, utilitarian structure to the venerable architectural styles of Scotland; the Edinburgh Old Town sixteenth-century model of desirable urban living, the gracious eighteenth-century New Town crescents, Glasgow's wally-closed tenements, all noise and society and thriving life. During late spring and summer, birds small and large rush in and out of the wall of leaves in an apparent harmony of ingress and egress, a many-storeyed dwelling-place, an example of multi-speciesism as blackbird takes its place above sparrow, blue-tit, thrush. The doves began their work of site management and construction (the first time I had ever seen any of them nesting outside their own house) in sunshine, squirming their well-fed bodies in and out of the ivy with a difficulty the smaller birds don't encounter, eventually laying and sitting, the fledglings hatching to bitter rain. I sensed them all season as a flickering, repetitive movement in the corner of my eye as I worked and they nested, tier upon tier in the ivy hierarchy, one tier up or one tier down, whatever this placement indicates, either the feral principles of the housing market or some unsuspected but elaborate system of avian social ascendancy that deems that blackbirds should be higher on the wall than sparrows or that doves, on account of some failure of timing or taste, be obliged to squeeze their way into the last available place, the least desirable, the one no one else wants.

Although I didn't quite see them I still became anxious if the rhythm of their coming and going stopped, or when a flurry of sound, disputatious or frightened, erupted from the garden. I watched but last year, as far as I could tell, there were no cats, no jackdaws, no magpies to disturb the progression towards fledging, no speculative raptors. After I'd been out, I'd come in to check to see that all was well – as if there might be anything I could do if it wasn't. The doves peered out from the ivy as they sat, calmly, implacably incubating and then feeding their young. As I always do, I wondered who was more aware of the fragility of the enterprise, them or me. Often I think of our shared vulnerability, the gales which seem to have increased, the heat, the rain, the debates about our one and only future that swirl in the air beyond this small, secluded realm.

I don't know why they nested there. Perhaps it was the observation of others, or the modish thing to do, the dove equivalent of a holiday home in Provence. It may have been the unusually warm weather during the nesting period. Perhaps it led to a proliferation of insects. (The rain began last year after the young had hatched, cold days interspersed with ones of unusual heat. This cold and dripping summer, they did not replicate their adventure.)

I watched all through the period of incubation. On hot days, I sat near them at the garden table, reading, working, waiting for the inimitable squeaking of a new-hatched bird.

There was one offspring. He has survived, but is confused. He was not there last night, in his usual place in the doo'cot. This morning, I couldn't see him. I imagined him strewn in his disassembled pieces, feather and disarticulated legs, stripped and disembodied feet, but there he was again, late in the morning, just outside the window, swinging on the clematis. I have studied him closely in the year since his hatching. He feeds with the others in the doo'cot but is last to observe the falling of dusk. In the increasingly chilly nights of autumn, with the change of the clocks, he stood alone on the roof at twilight, as if deciding between domains. If he has chosen to sleep in the ivy, in the morning he'll be standing on the windowsill above the study, waiting for me to open the doo'cot door for breakfast. As it grew colder, towards November, he began to file into the doo'cot with the others, but still was first to burst out when I opened the house in the morning. Unfailingly, he spends more time than any of the others near to his former nest, sitting often on one of the convoluted web of stems of the old clematis which hang like liana vines outside the study window, a Fragonard maiden on a swing. He seems to be always around the kitchen window, standing on the windowsill, looking in. This is his corner. Five yards away, his parents' home, a different place, not his. I watch as he struggles, as he pushes himself with some effort, trying to insinuate his adult dove body into the space where his nest was, trying perhaps to regain his nursery, to return to the state of the sublime perfection of babyhood to which we may all wish to return.

I wonder if Chicken is precipitate in her nesting, but she's not. Wherever there are rooks there is noise and industry. In the tall trees in Union Street Gardens, beside the city's main thoroughfare, the rooks' nests which stand all winter stark and black, exposed in the leafless trees, are now as frantic with avian activity as the streets and shops below are with the human sort. Every spring since I came to live in this city, I've watched and waited for the first rooks to fly back and forth over the lines of traffic, their beaks crammed with building materials, too busy to be bothered with drivers and shoppers and idle rook watchers like me.

The oddest thing about it all is that Chicken only began nest-building last year, changing the pattern of her spring behaviour from all the other years she's lived here. She has, in her own way, always acknowledged spring. I had become used to it, anticipating her running upstairs (usually in the early morning before anyone was around to stop her) to the first and then to the second floor, where she would stand for hours at a time, looking down through the banisters, a certain indication of spring. I never really understood why she did it. Was it a desire for height, a simulation of the experience of looking from a tree? I was torn between her needs and preservation of the stair carpet from the consequences of defecation. I tried compromise. I put down newspaper, which she ripped and scattered, and since I couldn't think of anything else that would prevent the inevitable consequences of her walking up and down the stairs, eventually, after many years of guilt and combat, of running upstairs after her, chasing and catching her before carrying her downstairs several times a day, I arranged a barrier of supreme inelegance, consisting of two large plastic boxes flanking an old lampshade, on the lowest tread of the stair. I did this only when psychological methods failed. I had first placed an item of which I knew her to be afraid – my embroidery frame on its stand, an object of enduring terror-inducement – on the wide fourth tread at the turn of the staircase, but spring determination overcame her fear.

It was last year when Chicken began her clattering, tugging nesting. I watched this new behaviour with interest and a little alarm, not knowing what impelled her to begin nesting in her later years. I wondered if it might be an intimation of mortality, a thought I have to accept but do not welcome. I looked around for information that might explain this delayed onset of proper corvid reproductive behaviour but couldn't find anything that mentioned the subject.

It began last year as it did this, with dish-clattering and then within a few days the sounds of newspaper-tearing, followed by the clicking of nails on floors as Chicken ran between study and kitchen, assembling nest-building materials. Her nest was constructed from newspaper from her floor, old receipts, magazine covers and anything else she could find or steal. I didn't pay sufficient attention to the process. It seemed odd but a better alternative to chasing her up and down stairs. The inconvenience of having a rook heaping torn newspaper under the dining table was limited. I did notice that her voice became louder. Talking to Bee or Ian on the phone, I'd say in explanation, 'Oh it's just Chicken in spring mode.' I watched her as she pottered about her nest, but her interest in it seemed intermittent and she continued to spend most of her time in the study with me.

Last year's nest was a haphazard affair, but the best possible one given the available materials. Even in the wild, rooks are known to

a nest constructed from newspaper

have a relaxed attitude towards the construction of nests, preferring loose, un-engineered structures, which sometimes fall apart and have to be reconstructed, the results it seems of their coming into the world equipped with more instinct than expertise, a situation improved apparently, as it may be for us all, by experience and age. (Other birds are different and seem to carry with them blueprints and scale drawings and a knowledge of architecture, the weavers and bower birds and house martins, consummate builders all.)

On the morning last spring when I went into the kitchen to see on the floor a small splash of yolk and a scatter of pale greeny-blue spotted shell, I peered dimly at it, failing to recognise immediately what it was or where it had come from. When I did at last realise, I was as astonished as Chicken was uninterested. She paid no attention as I removed the pieces of shell and cleaned up the egg. Before I had time to begin to phone the news around, I had found another egg, this time on the carpet, but nowhere near the nest. This one was intact, cold and equally ignored. I picked it up to keep.

As a proud new parent might, I phoned, sent photos of the egg to friends and family. Those long acquainted with Chicken were as amazed as I was. Congratulations were received, questions asked, mostly ones to which I had no answer: 'What took her so long?', 'Is that common?' For the people I told who didn't know Chicken, it seemed, reasonably enough, unremarkable – *Hey, guess what? Bird lays egg!* – but for us, apart from the shock, it was the first confirmation that we'd ever had that our original, chance decision to designate her female was correct.

This year, both Chicken and I are better prepared. We both know, more or less, what to expect. She has an air of authority as she goes about the business of construction. The nest grows as the locus of her existence and activity alters. This spring, she has moved from the favela house to the same place as last year, under the dining table, within the square bounded by the legs of a dining chair, a space that may seem to her secure, a dark suitable place to nest amid the items she's collecting up: one grey angora sock, a couple of floor cloths, the small, heart-shaped cover of a hand-warmer I bought last winter, a loose circle of torn newspaper. This year, her posture and behaviour changes even more markedly than last year. She squirms on the carpet, flattens herself, wriggles, lies still. I have never seen her do this before, legs tucked underneath her body, flat on the ground. She rests her head on the strut of the chair.

Her voice too has changed. Even louder than last year, it plays on a feral, manic note and now she sounds like an outdoor rook, one of the shouting, rasping confraternity of frantic nest builders, the wild rearers of young who even now have begun to haze the tree-tops with activity and noise. The timbre of her voice is rough-edged, ragged. I speak to her but she has an air of distraction. She is too busy for me. She runs with her urgent, hopping, rookish run between study and kitchen, with a combined air of busyness and necessity. She has no time for pleasantries. I sit down on the floor beside her. She comes close to me and shouts and snaps her beak towards me but she appears to have no time for me. When I try to tidy the edges of her newspaper circle, she runs at the broom — that object of fear and hatred — with which now, emboldened by some hormone storm, she is fearlessly willing to engage in combat.

In a letter to a friend, Gilbert White wrote: 'Rooks in the breeding season, attempt sometimes, in the gaiety of their hearts, to sing but with no great success,' and while I regret Gilbert White's lack of appreciation of rooks' voices, in Chicken I observe neither gaiety nor song. Fervid concentration seems to preclude both. For brief moments only she slips from spring mode, as if she has forgotten. Once or twice she jumps onto the armchair near my desk as I work, to stand beside me, climbing onto a cushion to be nearer still, but the familiar aspect passes and all too soon she returns to running, shouting and lying under the table, wholly engaged in the distant, mysterious activities in which I can play no part. This time, it is like living with a stranger.

The nest this year is larger than the last one. It is, as far as I can tell, occupied by Chicken alone, twigs and feathers and mud and all the other materials used by the outdoor rook being found only rarely inside houses. In this at least, I'm glad. The examination of a rook's nest described by Franklin Coombs in his book *Crows* lists, almost incredibly, among pails-full of dry sticks, grass roots, damp leaves, pieces of eggshell, potatoes and seeds, an astonishing population of creatures:

The live animal contents were about 50 earthworms; many brachyletrous beetles; dozens of nematodes of four species; hundreds of wood-lice, one slug (*Limax flavus*); collembolla of genera

Entombra, Lepidocyrtus, Orchsella, Pogonognathellus; a few dipter-
ous larvae and pupae; several spiders; a millipede (*Blaniulus
guttulatus*); acaris of genera *Nanorchestes* (20), Oribatid (1),
Parasatid (1) and a flea (*Dasypsyllus gallimulae*).

I don't know who I feel more sorry for, the rooks or the flea.

When the nest is a rough circle of eighteen inches or so, Chicken
begins to spend more time sitting in the middle of it. She begins to
caw and beg, loud and needy, as she would do if she were dependent
on a male to bring her food. Female corvids are fed by their mates and
even by 'helpers' during the entire period of preparation for, and eventual laying of, eggs. I begin to hand-feed her.

Just into March, on a Monday, towards lunchtime, she lays her
first egg of this season. (One study of the egg-laying timing of rooks
puts the mean date of the laying of eggs between 9 and 23 March.
She's correct. Again, I wonder how she knows.) The egg is broken
when I find it. She has been examining it but has already lost interest
so that, when I pick up the two pieces, she doesn't give any indication
of either minding or objecting. The egg is small, four centimetres
long, of a pale, delicate turquoise, a spray of pointillist chocolate-
coloured speckles concentrated towards the pointed end. The contents
are soaking into the rug. I clean them up but she doesn't pay any
attention. I sit down cross-legged on the floor beside her and she
jumps onto my knee. She has developed a 'brood patch', the bare
patch on her belly which in wild birds is in constant, warm contact
with the eggs. She gazes at me distractedly, then shouts. Her way of

communication has changed, her beak open wider, at a different angle, in the shouting.

I begin to think about her diet, that probably she needs more calcium, more vitamins to deal with the physical demands of the season. I consider what more I can give her. Her diet is as fine, as varied, as considered as any rook's in Scotland. Every morning I sprinkle vitamin drops onto her breakfast of cereal and milk. Is that enough? I take out an egg for her from the fridge. By comparison with her egg, the hen's is huge and pale. I break it and feed her some yolk and when I hold the spoon slightly above her head, she does what we have always called 'baby rook', simulating infant behaviour, crouching, opening her beak, flapping her wings with the gargling sound made by infant rooks as I spoon the yolk into her beak. I crush some shell into powder and give her that too. She settles again in her spot under the table. She is quiet again. When I come home, there is no voice to welcome me.

For many days, the behaviour continues. This time, I'm moved by the steadfastness she displays, saddened by knowing that she is dedicated to a task that can have no positive conclusion.

In mid-March there are days of bitter cold, although the forecast snow holds itself back, keeps itself inside the thin layer of dull grey cloud which is there and then is gone, melted into deceptive sunlight. I have to drive to Edinburgh through brilliant, cold sunshine and intermittent sleet. Coming back, the road north is closed after Dundee because three gigantic lorries have collided, killing one of the drivers. The police stand by the road to direct us all away, back the way we came. I travel north by way of the coast road, through Arbroath,

Montrose. It's familiar country, Montrose the nearest town to the village where David's grandmother lived until her death not quite two years ago, the town where we would come to shop, her anticipation of the pleasures of being among people and streets, in busy, lighted shops, undiminished in spite of her increasing frailty. Driving here makes me remember.

On this day, Montrose looks like the image of a Christmas town, etched and aquatinted, coloured by flinty, silvered light, dusted by frost, the country beyond opening into broad light, the sea glinting at my right. The bay, Montrose Basin, a bird sanctuary, is filled up with cold grey waves. Beyond the town, I drive over the river Esk, across the bridge where for years there was a slogan, now removed, scrawled by a patriot more daring than literate: SCOTLAND, FREE OR A DESSERT. This is the first time I have passed Gran's house since it was sold after her death. In the line of tall trees opposite what was her sitting-room window, the rooks are busy as ever they were each spring, the rooks she watched through the sights of an old Bofors gun, of provenance unknown, an endeavour that gave purpose and pleasure to her life, her morning and evening counting, her watching, listening, telling. Her house, when I pass, seems opened, builder's clutter in the garden, the windows wide and white.

Beyond, the road to Inverbervie is high above the sea, lined by tall trees, bent by the sea-winds towards the land. Fields stretch to the cliff-edges, small tractors balance dangerously on the edge of the world. Rooks fly overhead in black scatters, burdened with twigs, obeying the law that regulates their lives. A drift of them passes like dark smoke above the road as I pass the turning to Johnshaven. I know this road in every season. I have passed by at planting and growing and harvest, in every whim of the light, the sea's pewter glance or dazzling blind with sun, dulled over by fog and rain. I've driven here a hundred times as it has lost itself into darkness, the lights coming on in the villages one by one as I passed, the turnings to Gourdon, Fettercairn, Arbuthnott, Stonehaven, Muchalls.

When I get back, Chicken doesn't run towards the door when she hears me, calling, as usually she would. She is still lying flat under the table, surrounded by newspaper and dishes and the crumbs and scraps of every meal. Still sitting, she shouts at me. I hand-feed her as her mate would, as I did when she was a fledgling, putting mashed avocado and egg onto my little finger before thrusting it into her beak.

I worry about her, although I know that this mood, this need, will pass, as it does every year. Spring will shade into another season, one in which Chicken will resume her life, while outside the business of hatching and fledging is gathering momentum and with it the unimaginable dangers, the vulnerability and predation, when slow death or raw murder is everywhere, some of it brought about by the unthinking carelessness of man or the sport of felines, some of it part of the natural cycle of survival and death which may be difficult to witness and to understand but about which there is nothing we can do.

For the duration of the nesting, I abjure my usual cleaning, the vacuuming and carpet scrubbing necessary to ameliorate the tide of bird-related mess. I wash the floors but avoid Chicken's spring territory.

On Tuesday, I come back in mid-afternoon. There is another egg, perfect this time, unbroken, outside the ring of newspaper. She is not sitting on it and when I reach down to lift it, Chicken's irritated by the disturbance but not by the taking of her egg. She shows no inclination to sit on it. By Wednesday evening there is another. (Rooks lay eggs every twenty-four hours until the clutch is complete, beginning incubation almost immediately.) I don't know what to do, leave it for her in case she realises that she should be sitting on it, knowing that it will never hatch, or remove it before she stands on it. She's sitting, but not on it, so I take it away. It's cold in my hand, like porcelain. I put it onto a white plate.

By Thursday she's changing and I sense that the mood of spring, which has contained her so totally, is ebbing. When I get back from work in the mid-evening, she calls to me as soon as she hears the sound of the key in the door. She potters now between her nest and her house. Something feels different. I leave her nest alone for the moment, resisting the urge to tidy it away. When she comes to talk to me, her voice is no longer harsh.

It's as well that I leave the nest alone because her return to the usual recognisable rook is temporary. Within a day she's gone to sit under the table again, expanding her nest this time, tearing madly, leaving trails of newspaper between study and kitchen. She is no longer willing to return at night to the favela house. She is uncommunicative. She does not call early in the morning. I come downstairs to a small black shape on the carpet, to a grey face peeping from a cloud of newspaper, over the strut of the chair.

Over the next week, there are five more eggs. She sits, flattened to the ground, asking to be hand-fed, snoozing, head under wing. Again, she doesn't sit on her eggs as she should to warm and incubate them towards hatching. I leave them as they lie, cooling, until after a few days I pick them up. She knows, but she does not know. The behaviour is impelled by something beyond herself. Perhaps she knows they will not hatch. I put the eggs with the other one on the plate, a small, perfect still life. I would like to paint it but lack the skill to make it as beautiful as it is.

And then suddenly she changes again. It is Friday morning and with swift and comprehensive suddenness she seems bored by the sitting. She spends less time under the table and no longer snaps when I extend a hand towards the outer fringes of paper. She leaves her nest and when I begin, experimentally, to remove the paper, she doesn't try to defend it, doesn't shout and run at me. For a time she wanders between rooms. I remove the paper, a handful at a time. I wait between handfuls. She comes back and looks but is no longer concerned. I continue until it is cleared away, until I can vacuum the crumbs and leftover crusts of bread and butter, the scraps of dried avocado and egg yolk, and scrub the carpet. Chicken seems disorientated, purposeless, perhaps the way we all feel on returning to ordinary life after a removal from our quotidian routine, being unwell or away on holiday, even for a couple of days, when on returning we have, with unexpected care and attention, to reconstruct the small components of our lives. She follows me, keeping me close.

During the remaining days of March, the early ones of April, I watch her, sit with her on my knee, trying to judge what the season's activities have meant, whether she's changed, whether something of the spring has imprinted its losses, its deficits upon her. I do not know if I am able to tell.

Over the next few days, she returns to herself and to me. By the first evening, she has come to stand on my knee again. I look at her long, banded black feet and nails against the fabric of my jeans. Over the evening, she sinks lower, warming my knee as she does so, head under her wing. She is even more affectionate than usual, sitting very close to me, preening my hair, calling again from the bottom of the stairs. She comes to stand beside me as I work, jumping onto my foot under my desk.

One afternoon, I can't find her. I call and she calls back. She is behind a chair in the sitting room. She is bathing in the light from the sun through the front windows, wings spread, head to one side, with the deep and obvious pleasure birds derive from warmth and light. The preoccupations of sex and reproduction have yielded, abandoned their power over her. She is her own bird again, restored. Spring for Chicken is over. She seems content again to potter, to wander from room to room, to greet the morning with enthusiastic delight, the sun as it shines into the sitting room or onto her as she perches on her branch, to resume her bathing rituals and her evening habit of sitting on my shoulder as I read. The outer world is forgotten, extraneous, for this year at least, although it's still only March. For the rest of us, spring has barely begun.

A couple of days later, the clocks move forward. We lose our precious hour to cold, pale light. Chicken calls again, but now an hour earlier at the bottom of the stairs. From high ground above the city, everything seems wrapped around in chill, blue-grey mist. I look down on the canopies of trees, on dark smudges of nests among the branches, on rooks standing blackly, one by one, waiting, guarding.

14

Bird Music

Among the things of which we may be most afraid is the silence of birds. It foreshadows endings. It creates images of places despoiled, of habitats laid waste – oil slicks or poisoning by pesticide – of war and the threat of war, of destruction by heat or ice or water. The silence of birds is the absence of birds. Strip from each day, season, year, the presence of birds. Still the movement, stop the singing. In cities, the only natural sounds would be our own, our voices, the wind and rain. Birdsong is more than itself. It is not an addition, supplementary to the rest of our lives, pleasing but inessential; it is vital, necessary, for the sounds we hear daily are, at the very least, one marker, if not of our success then of our welcome failure, thus far, to complete our process of destruction. The title of Rachel Carson's startlingly innovative, seminal work on the effects of artificial pesticides on the environment, *Silent Spring*, published in 1962, shocks, for the terrible contradiction contained in those two words. An epigraph to the book quotes John Keats:

The sedge has withered from the lake,
And no birds sing.

There are very few places on earth where birds do not live. Even in the quiet, unpeopled regions we think of as remote, as high or cold or hot, there are birds. In the peerless cold of the Arctic, horned larks and ravens live, snow geese and owls, terns, rock ptarmigans, longspurs; in the Antarctic, penguins, skuas, terns, albatrosses, sheathbills, petrels. In the highest mountains, the Himalayas, the Altai, there are birds: the snowcocks, Tibetan and Himalayan, Caucasian and Caspian; golden eagles and lammergeyers, griffons and ibisbills. In deserts, there are plovers and nightjars, sand martins and sandgrouse, ravens and roadrunners. In every place, bird voices carry over distance on iced or burning air.

The song of birds is integral to our lives, one component of the synaesthesia of memory, like music or scent or the sudden image flashing a synaptic instant of recall, a summer garden, a wood, a moor, a harbour. A sound catches at you. You hear by chance from the radio the booming chanting of the ground-hornbill, its strange, almost human cry: 'Ing,' it intones, 'ing, ing, ing, ing,' and you are standing again on the central African morning, in the blue-gum-fragrant heat of another time. The recorded voice of terns surrounds you suddenly with image, sensation, memory: of rocks, sand, lichen, the crystalline emptiness of a Hebridean beach. The unfamiliarity of bird calls in another country can bring, as few other things, a realisation of the measure of distance.

As we sleep and as we wake, birds have slept and woken before us as the planet shades perpetually between darkness and light, latitude by latitude, the voices of birds rising and singing and ebbing back into night, like air-flows over the curvature of the earth, a never-ending curtain of moving sound.

Birds begin to sing at dawn, whether in the long, liminal polar dawn or the quick, brilliant dawns of the equator as the sky lightens before sunrise (the somnambulist's night- or daymare, the first notes of the crazy optimist's song from outside the window, the first intimation that attempts at sleep are pointless). It's thought that at least some of the purposes of their song, the establishment of territory and the finding of mates, are given additional force by the exuberance of dawn birdsong, a song often markedly different from one sung later in the day, carrying further through still morning air. Different species of birds begin singing at different moments, a stepped choir of evolutionary organisation. As dawn is still too dark for effective food-searching, birds may use the time to re-establish territorial rights, singing their most elaborate songs to impress and attract mates, among them new migrators who arrive at dawn. They may sing too in a post-darkness burst of enthusiasm, an expression of pleasure or an affirmation of rights, a bird's rare moment of dominion.

One year, when Radio 4 held a day-long celebration of spring, Dawn Chorus Day, we watched and listened as, one by one, tracing morning across the globe, bird and animal sounds animated each of our birds, lit in them, if not a cultural memory, then a spirit of interest or perhaps of celebration, as, in a song-fest of multi-species enthusiasm, they joined with great northern divers and trumpeter swans, with tui birds and musician wrens, with birds of paradise and marsh frogs, in inimitable song.

Roger Deakin writes enchantingly in *Wildwood* of sleeping out in a rook-inhabited wood on a warm May night, of settling to sleep after the rooks and waking at dawn to layers of birdsong, robin, chiff-chaff and blackbird singing under the overwhelming sound of rooks. 'I drifted back into consciousness to the most raucous of dawn choruses. It was still only ten past four.'

His descriptions of the sounds of beak and feathers, of mutterings and conversation, are as familiar to me as the sound of the voice at dawn. In this house, dawn, or at least the perception of dawn, has altered over the years, occurring now later than it used to. In her early years here, Chicken observed the rules of the natural world more strictly (as she still does on other matters), and dawn was dawn. It wasn't long after she came here that we learnt the daybreak singing habits of the rook, although we were lucky. She was never the earliest of her kind as she didn't call much before 4 a.m. and even then it was only on the light mornings of summer. Mostly, she called around five. We became used to it, the voice reaching us through layers of sleep. Now though, as she's grown older, Chicken likes to stay in bed, or rather on branch, later than she used to, particularly in winter, reluctant, as we all may be, to start the day in darkness. She'll respond to hearing the radio switching on upstairs by calling from her house as she wakes. After a short while though, bored by waiting for company, she'll come to stand at the bottom of the stairs as she's always done, calling to admonish or encourage. Her morning call is a regular rook 'caw', the sound we characterise as *aaarkh!* (our attempts at transliteration being primitive). Chicken reacts to what she regards as extreme laziness, or tardiness in coming down, with increasingly frequent calls, in a volley of *aaarkh*s from the bottom of the stairs. On meeting in the hall of a morning, we bow. She caws and I greet her. We bow again. She caws. I bow. She bows. I ask after her health. She caws. Eventually, we reach the kitchen.

Corvids belong to the order 'passeriformes' or 'perching birds', an order divided into two sub-orders, the tyranni (suboscines) and the passeri, or oscines, 'true songbirds', which comprise almost half of all bird species. Corvids are oscines, and though many people might question their inclusion on aesthetic grounds at least, that is where, by virtue of their evolution and taxonomy, they belong.

Only oscines, humans, whales, dolphins and some bats demonstrate vocal learning – the rest, anything else that utters sounds, does so because it is innate, not learned. Non-oscines such as doves and owls do not have to learn their songs; even when removed from other doves and owls, infants grow to make the sounds that membership of their particular species determines they should. Oscines on the other hand, like humans, have to be taught. Songbirds, parrots and hummingbirds learn their songs by listening to and imitating the songs of adult birds of their own species.

The songs of oscines are learned in infancy, most during the first two months or so of life, and although they may still learn later, their capacity to do so appears to be reduced. Studies have shown that if removed from their parents but taught song by a bird of their own species, they'll still learn to sing, whilst those who are not taught at all will sing abnormally. As with humans and speech, birds have to be able to hear their own song in order to sing. Young birds will 'practise', making small, indiscernible sounds at first, 'subsong', during resting half-sleep, an almost silent trill in a faintly moving throat, sounds that will develop and enrich and grow to become 'plastic song', which finds expression in the full, complicated songs of the adult bird.

The sounds all birds make are determined by their anatomy, of both their brains and their mechanisms of song production. At the base of a songbird's trachea, above its lungs, is that unique bird feature, the syrinx, by which sound is produced. There are many different syrinx structures, all complex, some super-complex, oscine syrinxes being more complex than those of sub-oscines. In most birds the syrinx divides into two cartilaginous rings, with two thin membranes – the tympaniform membranes – stretched between them, to enable the shape of the syrinx to be adjusted to create different sounds. On exhalation, the tympaniform membranes vibrate to produce sound, while the muscles of the syrinx alter the tension of the membrane to change pitch. In many birds, each side of the syrinx can function independently, allowing the development of intricate, layered song, sometimes two separate songs at the same time, in a bewildering variety, the vocal near-miracles that surround us in every shrub and tree.

(In the interests of research, an unfortunate crow was recently anaesthetised by researchers at the University of Utah and had an angioscope, the ultra-fine fibre-optic tube used to examine blood vessels in humans, inserted into its syrinx, which was activated by pressing on its chest. The mechanism that produced the resulting *caw* was captured, usefully enough, on high-speed video. Unpleasant though it might have been for the crow, it could have been worse. In *Il Libro dell'Arte*, the wonderfully informative fifteenth-century painter's manual, Cennino d'Andrea Cennini provides instructions for making casts of people, animals, fish and birds. The latter three, he helpfully suggests, have to be dead 'because they have neither the natural sense nor the rigidity to stand still and steady'.)

Birds have a set of brain nuclei and neural pathways, the 'song system' that controls both the learning and the production of song. The song system is made up of discrete brain areas – the High Vocal Centre or HVC, Area X and the Robust Nucleus of the Arcopallium, the RA. The song system is sexually dimorphic, larger in males, who sing better than females, the HVC of some male birds being eight times larger than that of females of the same species. In some birds the HVC and RA, subject to seasonal hormonal changes, are larger during the breeding season. Many parts of a bird's brain develop during the period of their learning song. In some birds the left and right sides of the brain can work independently of each other to produce a diversity of sounds.

The discovery of the process of neurogenesis, the development of new brain cells in parts of the avian song system, has excited particular neurobiological interest. Studies of avian brain areas responsible for song learning and song production, their mechanisms and the effects of damage to them on a bird's ability to learn or sing, appear to offer hope in the further understanding of similar structures and problems in humans.

Birdsong is a subject almost overwhelming in the range of its complexities and facts: numbers of birds, numbers of songs, each species with its varieties and variations, its responses, exchanges and dialogues, duets and mimicry, warnings and greetings, the messages we cannot yet understand. Birds sing, or make the sounds they do, for many of the same reasons that we use speech and song, to make ourselves known, to praise or warn or attract, for defence or for the expression of fear or feeling. The variations sung by individual birds are astonishing, as are the numbers of songs produced – the chaffinch may sing up to six songs, the nightingale two hundred, the brown thrasher two thousand. Hundreds of individual sounds may go together to form what we identify as one particular bird's characteristic song. Context and geography affect what is sung, through subtle alterations of frequency, pitch and tone. Assorted sea-birds calling together create a mad amalgam of dissonant sounds; the puffin, between complaining ghost and chainsaw, the high, persistent squeak of the Arctic tern, the petrel's rolling hiccup and the warbling choke of cormorants, the low resonant bassoon of the shag combine like an orchestra of maniacs equipped with instruments from a scrapyard.

Indeed, not all song is song. There are the clackings and tappings, the drummings and whistlings, the purposeful noises made by beaks and wings and feet.

Do birds enjoy their song? Does it only sound as if they sing, at least in part for the sharp, distinctive pleasures of exchanged speech, for the passionate, redemptive delight of raising their voices? The other evening, a spring evening of warm sunlight, the first in many weeks, I listened as Bardie and the blackbird in the ivy outside sang together, a douce duet, a cockatiel–blackbird fusion, a continent-spanning chorus of question and answer, phrase and response, prolonged and delightful, from which both parties, as it grew darker, seemed reluctant to withdraw.

On paper, birdsong can be represented visually by means of sono-grams, mysterious and wonderful, voices trapped on paper, graphs of time and frequency in infinite, beautiful patterns of line and curve and shadow, the song of any bird rendered in smudges of ink, repeat-ing and repeating.

The word 'songbird' itself is contentious. On morning radio a few years ago, I heard a woman in Glasgow talking about how she catches magpies *seriatim* in Larsen traps (which imprison the birds but do not kill them as gin traps would). She then takes the magpies out and kills them herself by smashing their heads against a wall. She recounted this with a degree of good cheer. Her actions caused her no distress. Her justification was that magpies attack songbirds in her garden. Her view is current, a continuing preoccupation, the belief that songbird populations are reduced by corvid – particularly magpie – depredation.

'Songbirds'. It's a word with form, a word that stands as an encomium to decency, to safety, familiarity, *niceness*, the word that springs forth from websites devoted to opinions on corvids or government consultations on wildlife control. *Songbirds*. Not to throw oneself behind campaigns to defend songbirds is to ally oneself with dark forces, to be, no doubt, in support of dangerous foreign powers, possibly subversive, to be un-British, wrong; the implied innate innocence, the assumed morality of songbirds, being set against the rampaging immorality of other sorts of bird, magpies with their flashy, vulgar black and white beauty, unable by sight or action to hide their evil desires, sensual, naughty, too raucous in their modus operandi, in their taking of eggs and young to pass by unnoticed as others do, the stealthy, the feline, the silent. Facts have never stood in the path of sentiment or ignorance. Magpies (or crows or jackdaws) do kill young birds but do not reduce 'songbird' numbers (cats, on the other hand, do). They are of course songbirds too, being oscines, of the sub-order passeri. They just aren't, it appears, the *right* sort of songbird.

Professor Tim Birkhead, among others who have studied magpies, demonstrates that the belief is wrong, that in fact populations of songbirds increase where there are also populations of magpies. Kevin McGowan of Cornell University suggests that, according to a concept he calls 'compensatory mortality' (using the inimitable metaphor of disabled spaces in supermarket car parks – if everyone could use them, someone else would get them first), if one predator doesn't get the young of garden birds, another will, that they'll be lost in one way or another, the balance of numbers being affected only when the natural levels of mortality are exceeded, the lives of young songbirds being brief, most dying within their first year of life. Reduction in garden-bird numbers happens where it does because of loss of habitats, through alterations in climate, pesticide use (including garden pesticides), the relentless building in cities, especially on what were once gardens, the increased numbers of cats, but not because of magpies. Every bird struggles to survive, to eat, to breed.

There is, I acknowledge, no pleasure in waiting for a pair of patient, watching jackdaws to take the young of the unfortunate dove who chooses, one year, to nest in the doorway of the doo'cot, but knowing that jackdaws' needs, as wild birds, supersede those of domesticated birds because their breeding is precarious, their prospects dangerous, their lifespan brief (doves can breed all year round, and often do, prodigiously; jackdaws do not), makes me accept nature as it is. Watching, as Bec did this spring from her window in Edinburgh, a magpie taking not only a clutch of black-bird's eggs, one by one, but the entire nest, to disappear high over the rooftops, might be a lesson in either regret or triumph but what it's not is a demonstration of evil. When the sparrow-hawk visits for plunder, I am philosophical. Sparrow-hawks must feed their young

a volley of aaarkhs

and they are, after all, part of a larger eco-system, which my doves, for all their qualities, are not.

I like songbirds. I nurture all the birds within my power to nurture, my own birds, the collared doves who visit the garden, the wrens, sparrows, thrushes, blue-tits, coal-tits, the blackbirds in the ivy. On cold summer mornings, early, I watch the blackbirds' two warm-brown-feathered young, looking strangely larger than their parents, in the first golden-syrup light before the city has woken, trying flight. The garden, overgrown, lush and green, organic, chemical-free, hanging with bird feeders, vibrates, hops, flutters, sings, alive with birds. The ivy wall sings. The shrubs sing. The roofs and chimneys caw and coo. The dawn choruses. Voices of blackbirds extend into the late, light northern city evening. From the darkness of their perches, the doves mutter and chant. I too wish for a pre-lapsarian world. I wish for the return of the Garden of Eden, perfection, innocence, for the abolition of death.

In the wild, Chicken's life would have been noisier than it is. Even the daily clamour of domestic life, the city assemblage of voices, music, doorbells, car engines, is dimmed by comparison with the sound of rooks in their natural habitat. Rooks, of all creatures, all birds, are accustomed to noise, both to making and to hearing it; social and socialised, they're used to crowds of hundreds or thousands, to jostling and shifting, squabbling for place, establishing rights, sleeping close, feather by feather, wing by wing. Their nights are spent among, between many others, the individuals who make up the vast clouds of drifting, settling, crepuscular flight, often a vibrant, raucous corvid mix, jackdaws, magpies, carrion crows, all huddling, cuddling together in high, swaying roosts, moving branches under their feet. They're used to numbers, to danger and the warnings of danger. They're used not only to noise but to surround-sound, high-volume, high-wire, high-amplification shouting and cawing, ultra-decibel noise, a hundred, a thousand orchestras, choirs, concerts of demanding, self-expressive noise.

Although to us they may not be easily differentiated, corvids have many songs and many sounds. They can recognise one another by voice, an advantage when males and females look the same. The question of whether or not their means of communication may be called 'language' depends on what 'language' is, but they appear to have what may be called 'syntax', the particular arrangement of sounds to form distinctive meaning. Some are able to differentiate between not only their own young and those of other birds, but individuals among their offspring. Repetitions, alterations in frequency and pitch, allow them a range of expression and communication, conveying meanings at which we can only guess. Corvid calls vary according to geographical area and may be specific to particular neighbourhoods as calls are developed, learned and exchanged within a group. They develop discernible regional dialects and accents. For all that, there appears to be mutual international understanding, for corvids in one part of the world, played the call of ones from another, will respond. If I play recordings of corvid voices from other places, Chicken will listen, bow and call, whilst the sounds of other birds (in particular the great northern divers, whose song I listen to frequently for the joy of hearing the pure, chill tones of northern melancholy) elicit mild interest but no vocal response.

Annie Dillard writes in *Pilgrim at Tinker Creek*: 'Their February squawks and naked chirps are fully fledged now, and long lyrics fly in the air . . . Today I watched and heard a wren, a sparrow, and the mockingbird singing. My brain started to trill why why why, what is the meaning meaning meaning?' We are in the midst of their sound, surrounded – as in a cosmopolitan city – by words of many languages we cannot understand.

It is not chance that made poets write of birds as participants in political or philosophical discourse; Chaucer and Henryson wrote of parliaments of birds, the Sufi poet Farid ud-Din Attar of a conference; for birds, particularly corvids, may have a disputatious, interested air. Crows together sound argumentative and vehement; rooks less angry, as if they're engaged in discussion and debate. Ravens' voices are rich and deep and almost human and seem to have a greater range of tone and expression than rooks' or crows'. Many corvids, particularly jays and magpies, are able mimics, both of other birds and of humans.

Attuned now to the changes in Chicken's voice, the range of her calls and expressions, I'm reminded of working, as I did in my youth, in the nursery of the kibbutz where I lived before I went to university, when I was a very junior assistant to the doyenne of all baby experts, a woman of profound and frightening knowledge who taught me, among other things, to distinguish between the cries of the many babies I was helping to look after. A woman of perfectionist rigour, she used to hold small, intensely serious seminars where I and anyone else in need of tutelage were gathered together. Frowning earnestly, we'd listen before opining solemnly on whether the cries we heard were of hunger or discomfort, distress or boredom. I did the best I could, given that, at the time, I knew even less of babies than I did then and later of birds. (Having worked all her adult life with babies, my mentor moved on in later years to manage what, under her care, became a prize-winning dairy herd, proving that knowledge may have unexpected applications.) I don't know if her influence prevailed but I knew when Chicken was hungry by the nature of her call. I began to recognise the obvious cries: hunger, impatience, alarm, great alarm, fear, mild annoyance, frustration, anger, the desire for company. Now, certain calls will make me run downstairs, make me run from my desk, the it's-a-cat! warning call, the sparrow-hawk alert (which will drive me into the garden with my plastic trident), the oh-God-it's-a-man-with-a-ladder call that signals the arrival of the window cleaner, a rather quieter occasion now than in the days when we had more birds, when the house would reverberate with assorted catamenial screams, rattlings, shriekings and flappings, as Mr Gordon and his assistant set up their ladders and began the terrifying business of wielding cloth against glass.

Chicken's sounds are diverse and subtle. There are the loud *aaaaarkh!* sounds of greeting, the *waaaa*, on a rising tone, like the

Chinese second tone, with which she expresses displeasure. There are subtleties of voice, murmurings and whisperings, the ones delightfully described by German biologist Eberhard Gwinner as 'tender whisperings', the ones in which the shadow, the echo of a word is there, 'hello' perhaps, inside the formation of a caw. People speak and write of the unpleasantness of corvids' voices but I don't agree. Chicken's voice is lovely. Infinitely, subtly changeable in both tone and volume, by season, mellifluous or strident, hushed or clamorously loud, it has the capacity to cheer or to rebuke. She speaks, I assume, the dialect of north-east Scotland, something like Lallans or Doric, except in this case north-east, or perhaps more precisely Deeside Rook. In *Wildwood* Roger Deakin writes: 'Rooks speak in the strongest of country burrs. They are rasping, leathery, parched, raucous, hoarse . . . and like all yokels, incomprehensible.' Yokel? Chicken? We cannot be thinking of the same bird.

Some years ago there was a cold afternoon, one of those days when northern light dwindles fast into early darkness, when even with the heating on the study takes all day to become warm. I was huddled in sweaters and shawls, sitting confined within the small glow of the light from my desk lamp, the cold blue stare from the screen of my computer. Behind me, from the semi-darkness, gradually, I became aware of a hint of sound, which grew, became the unmistakable sound of snoring. It was a murmur at first but grew slowly, becoming sonorous, loud, humanoid. There were only two of us in the study. I sat for a moment, transfixed. There were many things I didn't know about birds. One was that they can snore.

As I work, Chicken will approach my desk chair and utter a sound, a lowish *ehhhhhhhh*, drawing my attention to the fact she's there. I know anyway. I hear her toenails clicking across the wooden floor. I call, she calls. (I like this, the way she passes to and fro, the way the house is her own, the way she allows me to live in it.) Another of her sounds is one like old floorboards, old door hinges, old gates, a low, reverberant creak. While I'm speaking to one of the girls on the phone, they'll ask, 'Is that Chicken creaking?'

When I come back from wherever I've been, I unlock and open the outer door. From inside, beyond the inner door, I hear Chicken call greetings. I call back. Usually she is in the hall, or emerges from the study to greet me. If I've been away for a few hours or a few days, she'll run to meet me with wings outstretched, calling with what I like to believe is pleasure and welcome.

She comes, sometimes, when she is called. I, on the other hand, invariably do.

This morning I put on Jan Garbarek's *Rites*, music with the suggestion of birdsong behind the plangent saxophone, to encourage me to the pensive, post-weekend optimism required for work. Chicken begins immediately to ring, or rather to shake and tug at her bells in what is clearly protest. (I assume it's protest because if I turn the music off she'll stop ringing.) Perhaps today she's just not in the mood for plangent, melancholy saxophone music. This is another thing that I have discovered only gradually, that Chicken has preferences in music. Some pieces or composers she likes. Others she does not, in a range from mild to pathological. The latter degree of hatred she reserves for the work of Benjamin Britten. When, a couple of months ago, an excerpt from *Peter Grimes* was played on the radio, she uttered a loud and horrified squawk before running at speed from the kitchen, where it was playing. (This was not chance. When an excerpt from Britten's *Death in Venice* was played without warning last week, it evoked exactly the same response.) The Pogues' 'Fairytale of New York', heard all too frequently at the appropriate season, will send her from the room (although neither so swiftly nor so dramatically) and induce a frenzy of prolonged, irritated bell-ringing. Rautavaara does not receive even the courtesy of protest. At the first sound of 'Cantus Articus, Concerto for Birds and Orchestra', a recent gift from a friend, Chicken pointedly left the room to spend the rest of the evening in the kitchen standing under the table on a strut of a chair. Olivier Messaien, birds notwithstanding, receives similar treatment. She is, though, interested in Schubert. She likes Bach.

By chance, I happened the other day upon a particularly tetchy American website concerning corvids and their activities. There, I read the opinion that keeping any corvid in a non-natural situation will provoke madness in the bird, and whilst it may appear affectionate and happy, it is in fact unhappy, in a state of mental anguish.

For a morning, I worried about it. I did so on principle, because it is my obligation to worry, but also because that thought has, in all my many years of worrying about Chicken, been the one thing about which I haven't worried. It seemed to lead into a Freudian nightmare of denial, of wilful blindness to suffering. How might I know? How might I diagnose psychosis in her? Unhappiness? Considering the proposition, several things occurred to me. One was that the signs of unhappiness in birds and animals are usually quite obvious, in behaviour, appearance, demeanour. The other is that since it is difficult to find in any scientific literature anything that speaks in definitive terms of the cognitive or emotional life of birds or animals, to make such a firm psychiatric evaluation seems unwarrantedly bold. I looked at the website again. Then I looked at Chicken. The person compiling it was, I decided, quite wrong. If Chicken is mad, unhappy, anguished, then there seems little hope for the happiness of the rest of us, man or bird. Many things, I believe, tell me this and I can only hope that these things are true. Chicken's voice is one way I like to think that I can tell her mental state, the intonations and inflection, what I infer to be affection and intimacy, her pleasure in greeting and regret at parting. It is, I think, a contented bird to whom I say good-night, after the elaborate ablutions have been performed, after doors are locked, the house prepared for night, Chicken settled on her top perch. She calls to me and I call to her as we express our mutual wishes for the other's well-being, our hopes for the pleasant passing of the hours of darkness. We bow, as we did during our morning greetings. As I put off the light, I hear her scrape her beak once or twice along the wall of her house. In the morning, if I'm downstairs before she's up, I'll hear the low, soft, extended growl interspersed with high, soft cooing, the unexpected music with which a rook awakens.

15

Ravens

It's early April when my friend Chris, a dedicated naturalist, phones from Lochaber to tell me that he knows where ravens are nesting. Do I want to go to see the nest? I do. He says that he'll let me know when the young are nearly ready to fledge. Later in the day, I walk into town to buy a new pair of climbing boots. I've never seen a raven's nest but know that they favour high and difficult places for nesting and my only boots are old, the soles too worn, the laces too rotted, to allow me to reach wherever it might be.

Chicken has put her nesting activities behind her for the year. In these newly bright days of spring, she's more interested in seeking out patches of sunlight on the floor, moving from study to sitting room to kitchen, east to west, to find a place to sunbathe.

My new boots are whole and splendid. When I get home, I trog noisily around the house in them for a couple of hours before putting them aside until the ravens have raised their young. They'll have finished their building and nesting, flying with what they've found, what they've collected in the Lochaber hills, tangles of heather roots, strands of sheep wool caught on fences, other birds' feathers, small twigs from the stunted birches that grow beside falling streams. In previous years, Chris has seen young ravens about to fledge towards the middle of May, but this winter has been mild and they may have laid their eggs early. We don't want to go when they're still being brooded, in case we frighten them from their nest, or too late, when they've already gone.

I think that I have always wanted to go to see a raven's nest but I can't be certain. I can't remember what I felt before I knew corvids as I do. Now, I want to see these most intelligent of birds (possibly the most intelligent of all birds, not just of corvids), not simply from interest but from obligation too. I must because now they're almost family.

During the weeks when I'm waiting to hear from Chris again, there's enough spring around me to watch, sufficient birds. This is bird country, a place of cliffs and estuaries, flat land and mountains, fields and sea, the city like all northern, coastal cities loud with bird sounds, shadowed by patterns of flight, by mating and nesting, corvid caws and the insistent, rolling howls of herring gulls. High, wide north-eastern skies are suddenly alive with flight. Oystercatchers have begun nesting on flat roofs everywhere; hurrying, they draw trails of sound, their loud, insistent *peep, peep, peep*, across the city. In the gardens at Crathes Castle, a pair are nesting on the grass beside an ornamental pond. Gulls too are everywhere. In the city centre at night, they fly

270

through the bright white lights that flood city buildings, transforming them to luminous blue, lilac, silver against the darkness. They're nesting above the line of sight, on roofs, on chimneypots; the evidence of gulls is there too, on the tops of cars, windscreens, pavements, their voices in the air. I seem to hear them calling in my sleep and through my dreams so that, waking in darkness, they feel so near that they might be on the end of the bed, keening, calling down my ear with the voice of sea-ghosts, souls wailing from the fastnesses of oceans.

On every street, awkward, grey-brown, fuzzed infant gulls are staggering on knock-kneed legs into the paths of cars. One runs at my approach along a pavement, thrusting his head in skittering fear through the bars of a cast-iron gate. Judicious ushering allows me to manoeuvre him from gate, pavement and danger into the safety of the garden. I close the gate as his parents shriek from the roof above.

The annual complaints have begun about gulls' nesting, their ubiquity, their noise and mess. (One shits on my shoulder from height. In a small comparative study, I pay attention to the difference from corvid excrement. This substance has a gritty feel, a strong fish odour like very strong cod-liver oil.) A city-centre shopping arcade has hired the services of two hawks to keep the gulls from sharing the food of the al fresco diners (a rare enough experience in this climate anyway). Walking through town, I watch as a sandwich is whipped from the hands of a passing schoolgirl and swept towards the library roof gripped firmly in the yellow beak of a gull. She, fortunately, thinks it as funny as I do. Although I'm wary, as I always am in spring when the phone rings or the doorbell rings unexpectedly, fortunately no worried, kindly person seems to have come upon a small bird alone under bushes, assumed it to be abandoned, and rescued it.

In late April, I receive an e-mail from Chris warning me to be ready. A friend of his, another naturalist, has told him that the young should be flying by the end of the month. I take my boots out again and clump around, this time with a sense of purpose. When I take them off, Chicken removes one of my new climbing socks from inside the top of my boot where I have stuffed it and runs away with it.

Chris phones back a few days later. He has been to reconnoitre and has found the nest. It's high and difficult to see, but he thinks that the young are ready to fledge. I must go immediately. Alarmingly, he asks if I'm fit. Since I don't know for what, I find it difficult to answer. These things, surely, are not absolute. I say that I think I am. 'There's gym-fit,' he says darkly, 'and hill-fit.'

I pack, make sure that I've redeemed the sock that Chicken stole, and leave for Lochaber on an afternoon more like July than the last day of April. It's a day of mellow warmth, of clear, unbroken sun, the kind of sun that makes Scotland deceptive, other than it is, makes it seem unalterably idyllic, as if it's always like this, green and gold and timeless. It's a day which makes me, and possibly everyone else, forget winter, forget many shades of grey, forget cold rain, cold air, ice, unrelieved late or early darkness; transient amnesia, as if memory can be subverted so easily, just by a day of sun.

I drive out of Aberdeen, through Deeside and Strathdon. It's absurdly perfect, film-set perfect; time-stopped, sun-dappled, quiet roads through ancient forests of oak and birch, Scots pine, aspen. Ahead of me pheasants skitter their vague, panicky way across the road. To the side lapwings flap broad, flickering wings over green fields. Oystercatchers potter and pick by rivers where waterfalls cascade in glittering ribbons. No cars pass.

Ravens are even more difficult to disentangle from superstition and myth than crows and magpies, for they seem still more firmly woven than crows into the foundations of human cultural exposition, into flood myths, Sumerian, Hebrew or Assyrian, or into the cryptic, fantastical stories that have always been the way we've tried to explain to ourselves the ineffable mysteries of our own existence. (It's no wonder they appear in the way they do, as creators, all-seeing, bold, considered. We humans appeared late, stumbled hapless into a world inhabited by these intelligent, striking, canny birds. No wonder some of us believed that they created it all. Perhaps they did.)

In every culture ravens fly out of the days before memory, out of Distant Time and Dreamtime, to infiltrate man's dream life, the psychological haunting places where birds and gods seem one. One-eyed Odin of the Nine Worlds (Norse god of a wide and apparently conflicting portfolio of concerns: poetry, war, wisdom and death) had two ravens as constant companions, spies and bringers of news, whose names, Hugin and Munin, are Old Norse for 'thought' and 'memory'. Ravens have been the inspiration for royal houses, have named constellations, led men to battle under standards bearing their likeness. Regarded as capricious

273

and all-powerful, ravens dominate the spiritual life of many of the cultures of North America, Alaska and the Pacific Northwest. They are respected and admired for their observable qualities as well as their spiritual ones, their shadows moving over the ice at top of the world, sacred, spiritually powerful, protective intermediaries between the spirit world and the world of the living. In her book *Ravensong*, Catherine Feher-Elston describes the tribal groups Tlingit, Haida, Kwakiutl and Koyukon as 'Raven's children'. She writes too of her interviews with the shaman Medicine Grizzly Bear, healer and academic, who tells her that Raven, partner to the Great Creator, is able to intercede for both living and dead, to call souls back from, or to help them cross into, the spirit world. He talks too of the power of Crow, a different power from that of Raven; a power, among other things, used to protect soldiers in wartime. He describes the ceremonies and prayers used to extend Crow's protection to family members conscripted to go to World War II and to Vietnam, and of the ceremonies in the sweat lodge when they returned, when Crow was called on to purify them after the terrible experiences of war. (The sweat lodge, a sort of spiritual sauna, heated by hot rocks or directly by fire, is an important place of prayer, meditation and ritual.)

The appearance of ravens in Western literature is sometimes less respectful. Charles Dickens in *Barnaby Rudge* obliges the fictional raven (modelled on his own pet raven) to utter phrases as searingly banal as were ever uttered by any raven, or indeed human, anywhere, especially one that shared a house with someone who should have known better: 'Never say die!', 'Hurrah!', 'Keep up your spirits!', 'Polly put the kettle on', 'I'm a devil.' In his descriptions, though, he demonstrates a close knowledge of corvids: 'he fluttered to the floor and went to Barnaby – not in a hop, or walk, or run, but in a pace like that of a very particular gentleman with exceedingly tight boots on, trying to walk fast over loose pebbles'. Edgar Allan Poe refers in his famous poem 'The Raven' to the eponymous bird rather unkindly, as 'ungainly fowl'. Pliny redresses the balance by describing the funeral of the raven who, after years of affably greeting the citizenry of Rome from the sanctuary of a shoemaker's shop, was killed in anger by a neighbouring shopkeeper (possibly because it despoiled his shoes). As can only be right in such cases, the culprit was lynched and the bird honoured: 'The draped bier was carried on the shoulders of two Ethiopians, preceded by a flautist; there were all kinds of floral tributes along the way to the pyre which had been constructed on the right hand of the Appian Way . . .'

The road steepens towards Corgarff. The journey is one with built-in drama, for to take the road over the Cairngorms is to traverse worlds (admittedly, in a car, this cosmic matter can occur without much difficulty), to cross the vast massif, the mountains, Beinn a Bhùird, Ben MacDhui, Sgoran Dubh to the west and south, to be part of the high Arctic heart of Scotland.

Even the sight of snow poles lining the road, the snow gate standing open, redundant on this day of dark blue, sun-glinting roads, conjures the snow, although today the ski centre at the Lecht is brown and dry, the car park empty, the clanking paraphernalia of skiing sitting in idle silence. The road bends and curves and climbs. It's always scary, this road, although today snow and cloud and lowering rain are a memory of other journeys to the west, the ghost of the echo of us all singing one silly song or another fleeting in the air. The road climbs. *Engage low gear now*, and it's almost dizzying, driving round and down the steep bends of a road that is like a doorway in time until suddenly it descends into another country. The east is behind me. It feels so immediate, over the Lecht and then in so quickly into Speyside, Granton on Spey, Aviemore, so quickly the signs for Newtonmore and Laggan, the signs for the west.

Driving into Lochaber, I feel as if I'm home, that shifting concept. I've lived in the east for more years than in the west but feel still as a bird might, turned by sun or star compass, by magnetic force, a sense of orientation within, lodged, like a pigeon's, in the brain. (In the east, I always feel that the sea should be on the other side.) The weather here too is perfect after a winter of rain.

I remember that it's Beltane eve, the last day of April, one of the divisions of the old Celtic year. A perfect Bealtuinn moon hangs in the deep blue velvet sky. Tonight, Beltane processions will wind up Calton Hill, one of Edinburgh's volcanic hills. Fires will be lit; the May Queen will take her ceremonial place amid the Red Men and the Green Men, the host of revellers who congregate to celebrate the coming of spring. (If you didn't know otherwise, you might think from the web photos that it was a celebration of painted men in underpants.)

high and difficult places

In the morning Chris and I drive out of town, round Loch Linnhe to Loch Eil. We leave the car at the head of Glean Sron a' Chreagain. Most of the walking is on tussocky grass and heather, rising, steeply sometimes, over streams undercutting turf, spilling and seeping into dark peat mud, our boots squelching blackly, loudly, wetly, gravity's dark pull tugging at our soles.

A cuckoo begins to sing not long after we set off, accompanying us, seeming to be first to our left and then to our right. Whether it's the same cuckoo or many, I don't know, but the sound is with us for a long time, abandoning us only as we climb higher, when it goes silent. Then it's the breathless, high silence of hills, all sound sucked into the heather, all but our own breath, the sounds of our feet. It's too early in the year and too cool for midges, the ever-nagging bane, the ever-biting presence that drives some to fury and others to madness, and many to both. It could be, as Lochaber often is, wrapped into cloud and rain but today it's not. Pale butterflies flicker ahead of us among the first flowers. Tiny frogs dance at the edges of burns. Tadpoles swarm like dark spermatozoa under a microscope at the margins of standing pools. The occasional lizard darts from the approach of our boots.

The hill is Stob Coire a' Chearcaill, a 'Corbett', a mountain of 2,500–3,000 feet. The walk is more long than difficult but we don't hurry. It's quiet too. We see no one else all day. Not many people, Chris says, come this way.

Today is Laitha Buidhe Bealtain, the bright day of Beltane, the Gaelic day of celebration when Beltane bannocks, oatcakes, were once

baked and scattered in an act of propitiation to the forces of nature –
fox, wolf, raven, eagle – to ask them to spare sheep, cattle and chick-
ens their malign attentions. On this day ancient Celts purified
themselves with fires of juniper, celebrated the end of hibernation, the
new beginning of the fecundity of warmth and light. Beacons were lit
to celebrate the end of winter. Hearth fires were extinguished before
being ritually relit, and cattle were led between two fires in an act of
purification. The icy reign of the Cailleach Bheur, the Queen of
Winter, was ended, the power of Bride, the Queen of Spring, renewed.

Bride, daughter of Morrigh and Dagda, of the ancient fairy race of
Ireland, has her origins in pre-Christian, Celtic mythology, her iden-
tity merging with the first-century St Brigit of Kildare, miracle worker,
midwife, healer. She has two birds associated with her name, the raven
and the oystercatcher, the first because her feast day falls at the time of
their nesting. A poem from the Uists is reproduced in *Carmina
Gadelica*, a compendium of Gaelic hymns, stories and poems collected in the latter decades of the nineteenth century:

> *On the Feast Day of beautiful Bride,*
> *The flocks are counted on the moor*
> *The raven goes to prepare the nest,*
> *And again goes the rook . . .*

Up here, it's easy to imagine the fragility of existence, the need to
propitiate and ask and hope. There are fewer ravens now in Scotland,
and no wolves. The last wolf was killed in the middle of the eighteenth

century, possibly to the detriment of the raven population. The state of mutual co-operation between ravens and wolves is well established. Ravens both follow hunting wolves and lead wolves to carcasses, a process of benefit for wolves and even more so for ravens because, being unable to tear open animal skin themselves in order to feed, ravens need an animal that can initiate the feeding. (There has been discussion for years among wolf enthusiasts about the desirability of reintroducing wolves to Scotland. I like the thought. Apparently, they'd keep down a burgeoning red-deer population. They'd eat sheep too, but the rest of us would be safe since hill walkers don't seem to be a feature of their diet.)

A deer stands bathing in the water of a small lochan. It watches us for a while, steps out before we approach too near and lopes away. Loch Sheil's beyond us, scattering white light to the horizon.

We stop for a rest and as we sit an object, a black triangle, flies at speed down the length of Loch Linnhe below us, low, sharp, black. When we lived here, I used to rage as earlier models of this black triangle of death roared down the loch, too near to my windows and my children. I used to phone complaints to their base, sounding like the whining pacifist they must have thought I was, but I'd lived in a warring country and didn't want my children raised within the sights or sounds of war. They have to practise, the weary voice on the end of the phone always told me. Practise for what? And here they are, still practising, decades on, taking with them on their flight down the loch my naïve belief of long ago that international political progress is ineluctable, that things improve and justice prevails.

It's after we've been walking for five hours or so that Chris points suddenly to something I can't see at first, even through binoculars. It's a raven, high on the hill above, watching us, tramontane intruders into this silent land. There's no sound, then a single *orrkh!* In a quick, black flicker, it flies off. We walk on and hear the nest before we see it. From behind a buttress of rock, low sounds, muttering, croaking, reach us on the wind. The nest is in a cave gouged from a wall of rock at right-angles to where we stand and not easily seen. Below us the descent to the valley is steep, precipitous, not quite vertical. I'm secured by the rope Chris has brought. Chris isn't, which is quite reasonable. He is easy and brave with heights while I am a tremulous wimp. We have to edge down the grass and stones carefully, singly, to a ledge where it's possible – just – to sit. I edge along the narrow shelf of grass and lean out. Through binoculars, I find the nest. Below, the adult ravens stream round the space below us, two thin, crossing black lines.

Inside the cave, in semi-darkness, we can just see three or perhaps four grey shapes shift and move. In the five or six weeks since hatching, they'll have grown from pink, blind nestlings to feathered creatures weighing three or four pounds. (How much food-finding must that take? How much flying and looking and thrusting down three or four constantly waiting throats?) The young call, their voices low and asking, not the usual high voices of infant birds. A fold, like cloth, dangles a corner from the edge of rock. It wasn't there, Chris says, when he was here last. It looks like a piece of deer skin, furred, like a rug thrown over the edge. Wind moves through the rocks with a sound of blown grass. Stacks of cracked rock and layered grass are piled like an ancient desert city.

The site of the nest is majestic, the pinnacle of all nest sites, inaccessible, impossible to reach except by flight. It looks as if it has been used for a long time. Hanging streamers of grass trail far down the rock-face under the cave, whitened, stiffened. Splashes and runnels of white, thickly coated, decorate the rock, above and below. Above the cave there's a shape marked out in white. We discuss what shape it is. A sporran, Chris suggests. I think it's a shield. But it's both. It's a sporran-shaped shield, a raven coat of arms inscribed above the cleft of rock, ancestral, armorial. An escutcheon. Edging down the slope, we take turns to watch and in turn are watched. The parents fly and land and fly again, keeping us always in their sight. Inside the nest there's movement, a beating of wings, a dialogue of croaking. The young look as if they're preparing to fly, flapping their wings in practice. We sit for a while, take turns to edge down to watch. The infants are large enough to be alone for extended periods. Soon they'll fly. They'll stay with their parents during the summer, until they've learned what they need to know, most importantly to find their own food. We don't want to disturb them for too long. I don't want to turn and leave them but we don't want to keep their parents from them.

We walk higher, to a large, carefully constructed cairn. There are raven pellets balanced on rocks, grey-white, dry and ashy, constituted from greyish feathers, from pale slivers of bone, matted, felted fur, earth, raven spit. We pick them up and put them in our rucksacks with pieces of quartzite from the hill.

Driving home the next day, I think incessantly about the birds, wondering if they've flown yet and how, remembering the fragility of the lives of wild birds, the few that survive their first year. As I drive, the sun is beginning to break apart towards the east, clouds massing in a grey-blue sky. These days will be the last prolonged sun we have for months.

As I unlock the front door, Chicken runs from her patch of sunlight in the sitting room, calling to welcome me. I look into the doo'cot and replace a couple of new-laid eggs with cold ones. I check on the blackbirds in the ivy. Wood pigeons and collared doves are feeding at the bird-table. I put the pellets and quartzite on the mantelpiece. Over the next weeks, the pellets dry out, the grey feathers become white, and I wonder whose they were. The rain starts and doesn't stop, the beginning of another summer.

16

Of Flight and Feathers

On a July afternoon, I drive to the farm shop, a few miles from Aberdeen. Just before the turning to the farm, a black bird appears in the air in front of me, in slow, considered flight. It stops in the air, not hovering, not moving. Amazingly, it seems for this moment utterly still, suspended in the air, like a photograph, as if the world has stopped, as if we have become stone, or an image on a page. It is a rook. Slowly, assuredly, it floats into a turn, to descend towards something on the road in front of the car, holding its wings out, touches delicately down. There are no cars behind me so I sit while it redeems its unseen prize from the tarmac and carries it to safety on the grass verge. I have never seen this before, this stillness, this ability to hold the air, to freeze the frame. A rook. I am impressed.

It's universally acknowledged, one of those things we no longer even need to think about, that humans, or most of them, nurture the desire to fly, that everyone, sane or otherwise, must at some moment have envied the power of flight. No one, surely, can watch a bird step easily from the edge of a roof into that pure moment, to expand into the air over the sorry world below, without envy, without the shadow of the thought that it's not fair, that we have (albeit without much effort on our part) evolved to the high state in which we believe ourselves to be but still cannot, and will never, fly. Most of us accept with grudging equanimity that birds can and we can't and we're grateful (or not) for the little we can achieve, flapping a strapped-on, bound-to-fail accessory as we're pulled by gravity towards the bottom of the sea, or gazing down on cloud and sea and city from ill-ventilated metal tubes stuffed with people and trolleys of food and screens busy with entertainment to take our minds from the boredom, paradoxically, of flight.

There are those who can't accept it, as there have always been, those who will try anything to raise themselves beyond the restricting bounds of the earth. So far, every human attempt, mythological or not, to emulate the flight of birds has been risible, a history of crazy daring, of failing and falling, of everything complicated, risky, doomed, the true antithesis of the delicacy, lightness, the unmediated facility of bird flight, from Icarus's melting wings to the jet pack, all so very far from that one ethereal, lifting moment we will never achieve. Leonardo da Vinci famously said, 'For once you have tasted flight you will walk the earth with your eyes turned skywards, for there you have been and there you will long to return,' a sentiment that presupposes he knew from experience – which he didn't, because, works of genius though his drawings and inventions were, all evidence suggests that they didn't fly, or, if they did, not sufficiently well for him to be able to opine in quite such elevated tone on the pleasures of the sky.

There is a truism, ascribed to Churchill, that says that dogs look up to you and cats look down on you, while pigs look at you as equals. He didn't mention birds. Birds, I've learned, look at you from a measured distance. Birds, some at least, know that they know more than you. *We* – their look says – *we, apart from everything else, apart from our grievously underestimated intelligence, our so-far-unstudied tendency towards passion, our unexpected tempers, the ferociousness of our determination to please ourselves, we,* their calm, superior look says, *can* fly. In their look is the certain knowledge that they have uncovered the sharpest point of human envy, that one certain thing they can do that we cannot.

They make it look so simple. One summer morning as I walk up the lane, I watch a crow flying above the stone walls of the gardens. Last night's storm has calmed but it's still blustery. The crow turns sideways to the wind, seems to lie on the air that carries him forward. He turns, dips into a current, allows himself to be swept upwards, his wings held wide, black serrated feathers clear against the sky. His movements seem all pleasure, the sublime ease of his progress making me forget what flight is, what it entails, the long, slow evolutionary processes it has taken to create this melding of wing, feather and air, the aerodynamic complexities, the balances of force and lift, thrust and drag, the historic triumph this morning journey represents of creature over gravity.

Wings, most evolutionary biologists seem to agree, are exaptations, physical features that, having evolved for one purpose, evolved further for a different purpose entirely. To fly, you need a strong skeleton, rigid but light and hollow-boned, a keeled sternum, a furcula, joints that fold and lock, large, powerful flight muscles (pectoralis and supracoracoideus), muscles that pass through a hole in the bone, and the foramen triosseum, where scapula, humerus and coracoid meet to form a pulley system that allows the wings to lift. As it flies, the bird's furcula bends outwards to each side on the downstroke of the wings, recoiling powerfully on the upstroke. The bones of the wing are like human arm bones, but with fused wrist bones and only three 'fingers'. When Louis J. Halle writes of wings in *The Appreciation of Birds*, 'Man has never invented anything at once so strong and so delicate, anything so sensitive in its adjustment to the continuously varying conditions it has to meet,' it's impossible, after a moment's consideration, a quick mental scan of every human invention one can recall at the time, to disagree.

Thinking of the evolution of wings makes me see, as nothing else does, how much birds are part of the earth, part of the air. They are as we have ceased to be: indivisible from the circumstances of their evolution, evolved for climate and terrain; changes in human habitation over time, migration and the loosening of ties with places of origin mean that our skin and eyes, the shapes of our bodies, tell us little about who we are and more of who we were. Wings, in all their variations, their fusion of form and function, are what they have to be, suited minutely to their purpose and their place, adapted in every way to the demands of flight and life. The nature and requirements of flight determine precisely the shape of the wing, that ever-altering, moving, shifting structure of fragility and strength. Wings in flight are never still; in hovering they hold, wait, balance, learn the air.

Wings are curved, airfoils, convex on top, concave below, tapering towards the end, bisecting the air, which in flight flows over the top surface of the wing more quickly than over the lower, increasing dynamic pressure, reducing static pressure, while on the lower surface of the wing both dynamic and static pressure remain the same. The difference in static pressure between upper and lower surfaces of the wing creates 'lift', the force that allows the bird to fly. Each type and shape of wing deals in a different way with the forces of the air, with drag and lift, thrust and gravity, with the topography of earth, each meets its relevant climatic circumstance, from the lightest breeze to tempest, from the demands of stillness, the invisible, fluid trans-figurations of thermal change. There is powered flight – flapping flight (the complexities of which almost defy explanation) – and non-powered flight, gliding and soaring. Gliding is flying without flapping the wings; soaring is flight for which the bird uses energy from the movements of the air to fly. In dynamic soaring, a bird will use air streams of differing speeds and velocity to gain height. Large sea-birds will fly between waves, using the power of the air to carry them upwards. Thermal soaring uses the rising heat of air thermals, while slope soaring uses the warm air deflected from hills or buildings. Flight is more than a mode of traversing distance. Flight forms understand-ing with the air, senses wind, temperature, height.

Dynamic soaring is undertaken with high-aspect-ratio wings (the ratio of length to breadth), the wings of albatross, gannet, gull: narrow, smooth-edged, unslotted, elongated along their proximal edge, wings to give slow lift, to carry these birds vast distances for prolonged periods over open ocean, borne by the necessary wind. Corvid wings allow manoeuvrability for birds whose terrain is among trees, in gardens, city landscapes, those who have to turn, to negotiate, to flee, as do the wings of small passerines, of thrushes, blackbirds, sparrows, of grouse too, elliptical, low-aspect-ratio wings. The high-lift wings of eagles, storks, owls and hawks have slotted feathers, gaps between the primaries which increase lift to give them the wide flight they require for static soaring, allowing them more easily to carry prey. Falcons, vertiginous vertical divers, like members of the families Apodidae and Hirundinidae, the swifts and swallows and martins, the screeching speed-demons of summer towns, have specialised high-speed wings, unslotted, high-aspect ratio, as the wings of terns, ducks and sandpipers. For the swifts and swallows, they're designed not only for speed but also for feeding on the wing, for long migration, the unimaginable journeys during which they traverse the world in answer to the unexplained, instinctual compulsion within. And high-speed the wings are. Peregrine falcons can dive at 180 miles an hour (there are reports of planes diving at speed being passed on the way by falcons), although precise speeds are difficult to measure, being dependent on the strength of the wind. The speed range of birds is astonishing, from a leisurely 15–18 mph for sparrows to 217 mph for some swifts. Their wingbeats too vary with purpose and with size: seventy beats per second for hummingbirds, one per second for vultures.

The wings of owls are shaped to give silence to their noctivagant flight, their feathers fimbriate – fringed on the margins of the outer vanes of the first two or three primary feathers – so that contact between the air and the leading edge of the wing is muffled. Fine velvety pile on the dorsal surfaces of the inner vanes of all wing feathers and some tail feathers achieves the same, allows the owl his soundless nocturnal hunting. (I always think of owls when I read the metaphysical poet Henry Vaughan's poem 'The Night', where he writes of 'God's silent searching flight'.) It is feathers that allow flight, infinitely specialised in shape, form and colour, adapted minutely for each purpose and circumstance.

In 1861, the perfect impression of a feather was found indented into the limestone of Solnhofen, an asymmetric, secondary wing feather, at the time detached from its owner, archaeopteryx, who was found shortly after. The significance of the find was in that it provided evidence that feathers existed in creatures dating from the Mesozoic era.

It was once thought that feathers evolved from scales, but it is believed now that, whilst they share origins, feathers evolved differently, from the placodes – embryonic epithelial cells – which may develop into scales or feathers. The finding of dinosaur fossils such as sinosauropteryx showed that some had early forerunners of feathers, simple hair-like structures, 'dino-fuzz'. From these findings it has been possible to trace the stages of feather evolution from dino-fuzz – which may have provided insulation, waterproofing or camouflage – to the fully developed feather with its central rachis, its interlocking barbs and barbules: the feathers that give birds flight.

Light though they are, only 5 to 10 per cent of the bird's total weight, they're still two to three times heavier than the skeleton of the bird that carries them, this coat of asymmetrical wing feathers, feathers of flight, primaries and secondaries, the lesser coverts, which first meet the air, the main coverts in their rows, the retrices and remiges of wing and tail. Complex in structure, in colour, in form (used by the proponents of intelligent design, those creationists 'lite', as examples of entities, like eyes, too complex to have been produced by the process of evolution), they are composed of keratin, formed around a central shaft (the rachis), each vane made up of interlocking barbs and barbules, a system of hooks and catches that allows the feather to form a uniform surface to move through air.

Each feather has its place, the adaptation of size and shape determined by its purpose. A bird's contour feathers, held in place by a special set of muscles, insulate, provide form, colour and camouflage; carefully arranged and aligned, they keep the bird dry, protect it against wind, allow it to fly. Down feathers, which underly the stronger outer feathers, are soft, their barbs and barbules loose to trap air, which make them the most effective insulating material in the natural world. There are also semiplumes, somewhere between contour and down feathers, whose purpose is to maintain the form of contour feathers; and filoplumes, which are straight, more like hairs than feathers, their purpose still disputed – they may play a role in assisting the bird to detect the direction of air-currents. The largest of the contour feathers are the remiges of the wing and retrices of the tail, the former attached to the wing bones, the latter by ligaments to the pygostyle.

The feathers of the outer wing are long, narrow on the leading edge, shaped to aid flight, slotted to reduce air turbulence, whilst those of the inner wing are shorter and more symmetrical. Tail feathers are used for steering and for balance, and for the often magnificent displays by which male birds attract and display.

Many birds have a preen gland, the uropygial gland at the base of the tail, which secretes a waxy oil that they spread over their plumage, while some, doves and herons among them, have powder-down feathers, which keep on growing, eventually disintegrating into the fine powder that helps their owners maintain their feather condition. When I'm in the doo'cot, I see the powder dispersing into air, hazing a fine coating over the water in which they bathe. It lies as a pearly film over the garden pond when, almost as the observance of a spring or summer rite, they march down the grass in procession to bob one by one among the water lilies, to the horror and alarm, no doubt, of the unfortunate frog. (Can this powder be a contributory factor in the disease known as 'pigeon fancier's lung'? For a time, I thought it must be the serial Woodbine-smoking I supposed was indulged in by pigeon fanciers of old, in the privacy of their well-kept lofts.)

The alula, or alular quills, usually three feathers projecting from the 'first finger' of the wing, are used for fine control, for hovering, slow flight, for flying at the steepest angles. A bird can alter the amount of lift by altering the angle between the wings and the oncoming airstream, the angle known as the angle of attack.

Feathers vary in strength and durability. Many pale or white seabirds have black-tipped wings, the dark feathers that, because of their melanin content, are stronger than other feathers, being more durable for birds whose journeys entail prolonged flights over oceans. The larger ones with long, narrow wings have no need for the breadth of wing that would give them height. The wings of an albatross may be eleven and a half feet from wingtip to wingtip, the vast span that will carry it on its non-stop circumnavigation of the earth.

I go to stand on the edge of the cliffs a few miles south of the city, at Fowlsheugh, to watch sea-birds fly. Fowlsheugh – 'the overhanging cliffs of birds' – a place where the smell reaches you first, of bird and wind and salt. High paths trail the cliff-edges, seem to wander to the centre of the sky. A hundred and twenty feet of vertiginous rock fall away from the next careless footstep. Everywhere nests star the sheer, dark walls, guillemots and puffins, razorbills and fulmars. Nests scatter the ledges, balance in niches, hang on tiny spurs sliced halfway to the sea. The sea sends up its dragging, background roar, sucked from the deep caves cut into the cliffs below. The voices of birds, young and adult, mew and wail and chant, kittiwakes, black-backed gulls and herring gulls, shags. Birds rise in sprays against the high, cool sky, explode from faces of sheer rock into the rainbow haze of air, each wing, each feather carrying them on their dizzying spirals. (It's now, looking down on to the backs of flying birds, that you feel the gods might give a little help, transform you, let you fly.) I watch their different wings, their different flights, the kittiwakes with their narrow wings, their irregular zig-zag flight, the slow rising of the herring gull.

part of the earth, part of the air

It's early July but still cold. When it isn't raining, a damp north-east haar swirls over the town, round spires, through the tops of trees, over the garden. We might have difficulty in recognising its credentials as summer, but of course Chicken knows. It's moulting time and so feathers are on my mind – or, more accurately, under my feet. It is difficult to tell, but to judge from Chicken it looks as though moulting is an unpleasant experience for birds. During the weeks it takes, she's hingy. (This Scots word perfectly describes her state of mind and body; listless, unenthusiastic, *hingy*.) Summer is the time when her rich plumage of winter and spring is shed, not all at once but serially, feather by feather.

Moulting is a necessary process. Feathers wear out and do not regenerate. After a year of hard wear, they become raggy, they thin and break and lose their colour. Birds tend to moult in summer, between nesting and migration, when the weather is still warm enough for them to survive. For most birds moulting is annual, although for some it is bi-annual, those who live in conditions that put the greatest climatic or environmental stress on their plumage, in great heat or among dense foliage. Some moult partially, losing only some of their plumage, others fully, when all the feathers are shed and regrown. It's a time when enormous hormonal changes and adaptations overtake birds, affecting their blood chemistry, their metabolism, their susceptibility to infection and disease. They lose 30 per cent of their dry weight and require 50 per cent more energy. Their flight can be limited by their loss of flight feathers, leaving them open to danger.

Not just Chicken but all the crows and rooks I pass appear similarly hingy. Patchy-looking corvids, partially grey, with scruffy, unkempt-looking neck ruffs and thinning wings, are to be seen skulking on every roadside, picking in a dispirited way in the grass of city parks. *Summer?* their stance and posture suggests, *I could do without summer.*

The doves too are moulting, but their moulting seems lesser. A few older ones lose neck feathers and look for a time like small, grizzled white turkey buzzards, but the rest shed mainly down feathers, which drift into the corners of the dove-house, become caught in the spiders' webs that coat the walls and corners under the nestboxes and perches. Inside the house, feathers are everywhere. Scraps of black feather scatter the study floor. Wisps of grey down drift over the favela-house floor and under the furniture. (Had I thought of it a long time ago, I might have collected them and achieved the unique feat of having a rook-down duvet.) Regularly, I pick up the long, strong remiges and retrices which Chicken has tugged angrily from her wings and tail before they fall out naturally. I use them as bookmarks. Pages bristle with them.

I pick up feathers and lay them in a line. Chicken's are not just black. They're tinctured sloe and ebony, ash grey, powder grey, charcoal, silvered navy. I read through a list of the names of the birds of the world and am dazzled by words of colour and brilliance, with a blaze of polychromatic multiformity: feathers, the meeting points of pigment and light, some of them pigments that exist nowhere else, the porphyrins and psittacins, although there are the carotenoids too, the melanins, the biochromes, which tint and paint each feather of every species of bird on earth. The words used within the names of birds are heavy with colour, with gemstones, jewels, metals, minerals, spices, words that can only partially describe: violet-green, lavender, lilac, cream and ivory, cinnamon and chestnut, apricot and tawny, pearled, bronzed, copper, golden, opal, silver, sapphire, topaz, emerald, fire. 'Fire-maned'. Words of *haute couture* too, employed to describe texture and form: crinkle-collared, velvet-mantled, striped, braided, spangled, forty-spotted, black-bibbed, crimson-hooded, the ways of granting a place to each parrot, each wren, tanager, weaver, sunbird, goose and hornbill.

The birds of our daily lives are as exquisite as any: starlings, which, close-to, look as if their plumage has been knitted on the finest needles from thread of drawn metal into verdigris and tortoiseshell, lit by flashes of light, shining black and brown, each pointed feather edged by a tiny vein of gold. Before I knew a magpie, I would have imagined its feathers to be simple, either black or white, but I would have been wrong. As Spike moulted during his first summer as an adult, the feathers he shed showed multiform gradations of black and white, feathers of white edged with black, of black edged with white, some of black bordered all round by white; all, from the largest contour feathers to the smallest, composed of the finest differences in the proportions of colour. He shed down feathers of soft grey, curled white feathers of pearlescent delicacy, long, gleaming wing and tail feathers of intricacy and perfection. Every year I'd gather them up in handfuls, wondering what I might do if, as in some insoluble fairy-tale task, I had to reassemble his plumage from this mass of super-complex, subtle coloration, calculating how long it would take, if it would take as long as it took Time to make him. His wings and tail were iridescent, granules of melanin reflecting, absorbing light, turning them crystalline blue shot through with shimmering turquoise. In flight, a magpie's wings are fans of glittering blue, edged by a fringe of white, a white cape, a Celtic torque of silver circling its back.

Like the colours, the names of birds mesmerise with a wild, descriptive poetry: spangled honeyeater, bronze-olive pygmy tyrant, fiery-tailed awlbill, forest elaenia, lovely cotinga, white-winged fairy wren, Bishop's oo. (The latter, a Hawaiian bird, is either critically endangered or actually extinct. A lobelia eater, sadly it has no eccentric ecclesiastical connections, having been named for a naturalist called Bishop.)

Familiarity doesn't dull me to the wonder of birds, what they are and what they do. Chicken becomes more mysterious, more miraculous the more I learn, the more I observe. I spread her wing in my hand. She grunts and, briefly, objects. Before she tugs it back under her own control, I look at the lovely arc of it; feel the fine bones under my fingers, feathers all in their symmetrical or asymmetrical orders. I think of the fact that she doesn't fly, another of the deficits of our long-enduring contract.

Apart from treading on a carpet of feathers, it's at moulting time that I get the true measure of birds, when I remember that they're all illusion, con-artists, feathers a magnificent, often ostentatiously successful cover for not very much at all. Feathers are costume, all keratin and mirrors, covering creatures that are, without exception, smaller than they seem, scrawny things under their blanket of insulated plumage. Spike, apparently plump, full-chested, dense, under the lush feathers was tiny. Moulting time revealed his neck, the diameter of my own little finger, denuded of its thick, marvellous glowing blue-black plumage, a pink, goose-pimpled, frankly alarming manifestation of his true size and fragility. A dramatic moulter, he shivered and cringed and drew his tiny pink dinosaur neck and the rounded, shockingly nude back of his head into himself. He hunched under the counter lights in the kitchen, standing on the jar of shells, glaring, seeming full of mordent self-pity. Every year I was scared that his feathers wouldn't grow back and that he'd proceed through life with a partially bald pink head and naked pink neck. Fortunately, it never happened. The pimples would expand, begin to form the small quills of pin-feathers which would break the skin, growing rapidly larger, unfurling from their keratin sheaths. As it does for Chicken every summer it kept him occupied, as, becoming refeathered, he grew in confidence, busy about the tidying-up of his feathers, picking off discarded casings, regurgitating them eventually in one of the revolting pellets he routinely disgorged from his beak with supreme thespian flourish.

Bardie too is moulting, the floor of his house feathered with tiny wisps of grey and white and orange, some so small that they seem too tiny for their own perfection. At this season the treats I buy for him from the pet shop are special sticks of seed called 'moulting-bars' which contain ingredients designed to ameliorate the torments of moulting, oats and honey and many, many vitamins. (It's difficult to tell if they work. Every year Bardie recovers, regrows his feathers, remains as hostile to me as he has always been.)

Marley the sun conure annually shed his feathers of citrine and saffron, of pure scarlet and orange, scattered fragments of brilliance cast on the floor of his house. I have two of his wing feathers in front of me on my desk. They're grey on the broader side of the rachis, pure leaf green on the other, shading towards the tip to a point of indigo and violet.

At this time of year, Chicken's left eye waters. It is an event that causes me anxiety every year. I bathe the eye but it continues to water. She sits on her perch quietly grumpy, waiting for the colder weather, sulking her way towards autumn when, magically, majestically, her feathers will grow back. After weeks, the day arrives when, stroking her head, I feel pin-feathers pushing up insistently from under her skin, through her remaining feathers. They begin the breakthrough, a slow, meticulous unfurling in this stunning progression that is the annual cycle of avian life. She is a busy bird, concentrating, removing feather sheaths, running her wings through the brisk pincers of her beak with a rush, preening and bathing. Then, autumn, and Chicken is beautiful again, restored, thickly, glossily slick, new, fresh, black glowing with a purple-navy-almost-golden glaze. Her neck is rich and full and warm. Her feathers smell sweet and new. Everything in her behaviour tells me that she feels restored. She walks with pride and, I think, with relief. She is no longer hingy. I buy a smart new black coat. Winter can arrive and together Chicken and I will be prepared. The other birds too will grow their feathers back and be ready again to taunt humanity, to remind us that, if we desire flight, we will never be free from the encumbrance of someone else's wings, that, until we find a way to evolve, by running to catch insects with our hands perhaps, leaping from tree to tree for a spare few million years, we will achieve nothing; whilst they will be there, always, to remind us that the creatures some so airily call 'flying vermin', the scruffiest street pigeon, the modest sparrow, can do the one thing that we cannot.

17

Ziki

Autumn now, the days of the autumnal equinox just past, a change in the air, the sky. It turns in a night from chill and raining summer to chill and raining autumn. There have been a few, rare days of autumn sunlight in between the rain. The city changes, green to bronze, copper, fire against the grey. Leaves have begun to drift like tawny rain. The skies will become migrants' skies. One morning we'll hear or see the geese returning. Some of us will be caching, as we do at this time of year, preparing, storing up for winter. Those of us with brain components that increase in autumn, whose hippocampuses expand to meet the challenge of the days, will be busy hiding, checking, remembering. Soon, the clocks will change. An hour will be regained at the price of early darkness.

This autumn we have a new crow. He has been here for some weeks now. He doesn't come from lovely, rural Deeside or from the heights of a monkey-puzzle tree in a quiet, well-kept garden. This bird comes from a 'backie', an urban back yard of scruffy grass, a few trees, near a busy roundabout and major roads of traffic and fumes and noise.

I'm in the pet shop when I learn of him. I'm there to buy a sack of dove food. 'Oh,' the assistant says on seeing me, 'there was a lady in about a crow.' She searches in a notebook and brings out a piece of paper. Handwritten, just an address and a name, and at the top, those two most enticing of words, *baby crow*. 'We told her you've got birds,' the assistant says. (It's the sacks of dove food, the moulting-sticks and grit, the bird vitamins, the feeding dishes, the cage toys, the wild-bird food, the egg biscuits, the doves I bought from them, that give me away.) I take the note. What am I to do? I should, at the very least, phone.

I do, and then I go to visit. I am, although uncertain about it all, more or less prepared. I have in the back of the car a box, half a dozen eggs, a packet of mince. (*Feeding Cage Birds* has been consulted again, the loose pages realigned, page 139, 'Crows, Jays and Magpies', relocated where it should be, between page 138, 'Broadbills and Apostlebirds', and page 141.) I hesitate because this bird may not be a crow. It may not be young, or it may be ill with an unknown avian malady that could infect the other birds.

I find the house and am ushered into the presence of the bird. He is a crow, and a young one, but he's not so much of a baby any more. He's feathered, sort of: scrappy black, unkempt, with scruffy, torn wing feathers. His face is extraordinarily beautiful, his eyes the largest, most searching I have seen among many beautiful, searching corvid eyes. His age, if I had to guess, would be three months. He has had a difficult life to date. I know no more than what I'm told by the lady who rescued him, an incomplete tale. He was being attacked, she said, by dogs, seagulls, by – and she said this with shock – 'his own kind'. One night, she gave a tenner to a drunk man to catch him.

For all that, he's attentive, bright and watching. When I visit, he runs from behind the sofa to look at me. I squat by the side and call him. I'm careful in the way I address him. He appears, watches me, runs away again. He comes back, interested, maintains distance between us. It's clear that his rescuer finds the situation difficult. (It is her first encounter with a corvid. She's amazed by how clever he is.) She has nurtured him, kept him alive since his rescue six weeks ago. She's tearful. She is parting with him only because she has to. I can't take him away so abruptly, put him in a box and leave. I say that I'll have to talk to my family, although I know what they'll say.

As soon as I get home, I begin to think, to assess the implications. Where will he go if I don't take him? Would his future be better any-where else? I think of his name. He has been named already: 'Beaky'. For many reasons, this will not do. For one, Bec is known by Bardie as 'Beek', shouted, sung, shrieked, 'Beek, Beek, Beek', a refrain we do not hesitate to purloin, and so the possibilities for confusion are too great. I think, read, study and come up with the shortened form of the name Yehezkiel, Ezekiel. It is Ziki, close enough to be familiar, but dif-ferent too. My feeling that the bird is a male is, as ever, beyond immediate substantiation, but he seems to me to be large for his age. He just seems, don't ask me why, male. The statistics seem to work. If every other bird is deemed male, then occasionally we may be right.

During the days when I'm planning, considering, I think of the bird and of the circumstances that cast him from the life he might have known, of the events in the lives of us all, bird, beast or human, that may seem like chance but aren't, the upheavals and cataclysms of politics or nature that dispossess or destroy, taking us all to other lives or other destinies, of the interdependence, the solace of others, on which we all may or may not depend. I think of how, after I first met Chicken, the feeling began to develop that, more than simply taking a new interest in corvids, I had opened myself to a new society. It was as it might have been to find oneself included in the ranks of a powerful secret organisation, a corvid Cosa Nostra perhaps, to have married foreign royalty, formed an important dynastic alliance. With it came the deeply reassuring thought that anywhere, more or less, where one might find oneself upon the earth there would be at least one familiar figure to offer greetings, familiarity, a sense of home.

And so it was. Travelling in Lithuania, once in a warm and muggy early summer, once in a freezing, snowing spring, I understood the power of the alliance. That summer, I spent time in some of the sad and terrible places of death that are the last war's legacy to Europe. In one of those places, in the woods at Panerai, only the sound of birdsong, some of a kind I had never heard before, connected me to a living world. In the flat in Vilnius I had rented, I'd open the windows in the early evening to hear the only sounds that could make me feel less alone, the sounds from the trees, of gargling, choking, squawking from between the bright leaves as young rooks were fed.

On the spring visit the cold was exceptional in its ferocity. Intermittent falls of April snow, a wet, discouraging snow, coated the medieval streets of Vilnius and fell into the Easter baskets of children walking to church. That time, I was not alone; Han was with me in retracing my steps of a few years before. On a dark Saturday afternoon we were the only people at Panerai, the only living people. On the margins of the thin forest, a railway track and falling snow.

Later, standing on the station platform, waiting for a train back to Vilnius, we froze and hopped and danced to try to keep warm and fed chocolate to the rooks who gathered around us, just a few of them, larger rooks than Chicken, but of the same familiar face, walk, stance. We supposed that sometime a train would come. The trains were high, with the look of a former age, a *Dr Zhivago* air. As we stood there I thought of Stalin's designation of Jews as 'rootless cosmopolitans' and it seemed, contrarily, a pleasing enough designation, one that might have done for either of us, human or bird, and by extension for many of us, sojourners on this soon-to-be-ruined planet, because all our lives are fissile, brittle, subject to contention and to storm. The rooks surrounded us companionably until the train came, eating chocolate, their black feet making the lightest indentations in the snow.

Three days after my visit, I am driving home with a large wire house taking up most of the back of the car, a cat-carrying box turned temporarily to crow-carrying box (with attendant crow), the sack of dove

a crow, and a young one

food I hastily bought on the way to collect him, a couple of bags of wild-bird food, some extra bird-feeding dishes. (The parting was anguished. I felt like a birdnapper.) There is no room to operate the gears. I lurch and grind past the High Street, up St Machar Drive, across Rosemount, negotiating roundabouts with the eccentric speed variations of tortoise and hare. It is one of the only warm days so far this summer, and will be one of the last. Summer, by now, is ending. Ziki and I steer eccentrically down the road in sunshine and are home. I carry in bird and bird-related items.

I set up the commodious house in the rat room. It puts the favela dwelling to shame. I cut some apple branches from the garden and secure them at different levels through the wire and position the cat carrier at the open door of his house. Ziki cowers in the cat carrier, coming out for meals when I leave him alone to eat. He eats interestedly, avidly – egg yolk, mince, brown bread which he drops into his water-dish to soften.

After two days, when Bec is visiting, together we remove the cat carrier. He plays peek-a-boo with Bec. She calls 'Hello' and 'Goodbye' from the door of the rat room and his small black head peers round to see her. He seems a bright, enquiring bird. His head is oddly decorated with dried-on food scraps, which make him look as if he has particularly matted, blond dreadlocks. It's difficult to avoid: small corvids are messy eaters. I still have the face cloth with which I used to try to clean Chicken. Their skin is delicate and cannot be rubbed with too much vigour.

One afternoon, we decide we must clean him. Bec reaches in and gently catches him. He, in return, snatches a pinch of the flesh of her arm in his beak. He closes his eyes and, rapt, he holds on. I have, carefully, to reach in to prise his beak open sufficiently to release Bec's arm. Freed, Bec wraps Ziki in a towel. We take him out of his house and I dab his head with warm water to prise the dried muck from his feathers. As I do, he becomes a crow again, not a small, corvine hippie. When we've finished, carefully we unroll the towel on the floor of his house. He lies quite still, eyes closed, unmoving. Has his heart stopped from fear? Has he stopped breathing? We look at one another as our hearts and breath stop too but slowly he opens his eyes. He looks round, stands up, shakes, then leaps onto his perch.

'God,' Bec says, 'they're such drama queens.'

The experience doesn't damage his appetite. He eats a large supper, raw mince, smoked trout from last night's dinner, cheese. He bathes and hops to and fro on his new apple-wood perches. I talk to him.

Ziki's house is on the floor on one side of the long, narrow rat room. By opening the door of his house and wedging it against the washing machine I can create a small enclosure for him at the far end, with a view through the glass door to the garden. He has space in which to roam and play and bathe. He doesn't play. I give him toys: a ball, some old toys of Spike's, a few small boxes, and – knowing Chicken's response to them – some bells. I thread some small, brightly coloured ones, red, purple, gold and green, on to string and hang them from his roof but he shows no interest in them. It is too early. I give him a large shallow dish of water. He bathes several times a day.

Chicken, who has watched him with increasing interest, can see him but cannot reach him. She has taken to standing on the strut of the dining chair from where she has the best view of Ziki and his house. She is watchful, bossy, aware, it might seem, of her status. I wonder how she sees him. She approaches his house sometimes with an aggressive posture, feathers fluffed, head down, and tries to snap at his tail. He looks at her with equanimity. I'm sure it won't be long before he asserts himself and responds to her aggression. I don't know what he will do.

Unusually, a couple of weeks ago, a large, adult crow began to appear almost daily to stand on top of the rat-room roof, to feed at the plate of bird food on the garden table, the first time I have ever seen a crow do this. As with Spike, I wonder if they see through windows, if, by some mechanism of watchful communication, they know that Ziki is here.

There's one thing that's odd about this bird. He has made no sound. He is completely silent. Even when being held, he made no sound, no shout, no normal loud, vehement corvid objection. There is something deeply unsettling about a silent bird. I search in all the available literature to find precedents, possible causes. I don't know if he can't speak or hasn't learned, if he might have possible damage to his anterior forebrain pathway, his HVC. If a lesion occurs during the process of learning song, the bird will not sing. It could be damage to any part of the complex neural pathways of his forebrain that makes him silent.

His right foot too seems less efficient at grasping the branch. We wonder if the two are connected. A bird's speech centre, like a human's, is on the left of its brain; thus, damage to the left of his brain might render him both speechless and lame of right foot. Might this be the reason he was rejected then persecuted?

If he remains silent, his future here is assured (as it probably is anyway). A voiceless bird has no future among other birds. A voiceless bird cannot communicate, cannot call alarm, cannot woo, cannot exchange the myriad social calls necessary for the safe functioning of corvid life. Spike stayed because he spoke. This one, almost certainly, will stay because he can't.

I think of his early experiences. Whenever I look at him, I realise that this will all take time.

Over the weeks Ziki grows more feathered. He is becoming glossy. I see him stand one morning on top of his house in sunshine, looking as if he is surrounded by a nimbus of gold. It is not prejudice that makes me admire the beauty of this bird's face. Everyone who has seen him comments on it too.

To test his hearing, I play some CDs of birdsong to him. The ones I have are of American birds. None of the assembled listeners, on a cold Monday morning in August – Ziki, Chicken, Bardie or myself – has heard any of these sounds before. Chicken is interested but unmoved, listening but not agitated. Bardie cheeps intermittently, as if in comment and agreement. Ziki, on the other hand, runs animatedly about the rat room, listening, responding, looking around, looking up, wondering, fascinated, bemused, more animated by some sounds than others. His hearing is obviously intact, but he's still silent. The sounds of mobbing birds appear to agitate him. He has Spike's watching, brilliant eyes and looks like the Chinese painting of a crow that I've been given recently. There are two characters written by the side of the crow which I translate, brilliantly, as 'young crow'.

His body is fluid, mobile, different from both Spike's and Chicken's. He is like a small black seal. He hops in his house, corner to corner, corner to corner. It worries me. Again and again, corner to corner.

One morning, we listen together to a Welsh mezzo soprano singing loudly from the radio. Ziki is on his perch. He raises his head, his face a study of rapt, intent delight. (I am sure I am not mistaken in this interpretation. He seems disappointed when it ends.) Later that morning, although it's still far too early, I begin writing my Christmas list. Ziki's CD is the first and only item. I begin to put on Radio 3 for him to listen to when I'm away from him. I check to see who is 'composer of the week'. He seems to like high sounds, early music, the human voice.

Then one morning I hear him playing with his string of bells. He spends some time shaking them. I sit next to him and talk. I haven't got the time just now but I will. I will spend more time with him. I will sit beside him. I will talk to him. I will wear him down, refuse to accept his fear. I will send him the message, somehow: *This, boy, is all there is.* He has to know. No one will be more interested in his future. He will adjust. It will take time. He will come through to the kitchen. He will accustom himself. One day, inadvertently, he will jump onto my foot, surprised that he has done so. We will talk, even if the conversation is one-sided. I am confident, determined. There is nothing else that can be done. As with the others, I see him thinking although he doesn't speak. I'm sad for him, for what must be the background of his life; I think of it: a young bird, a drunk man, the darkness.

The crow is there again this morning, outside, first on the roof and then on the table, his feathers richly blue and purple and black among the maple leaves that are falling from the tree, still brilliant red as they fall. I don't know if Ziki sees him, if he does, what he thinks or feels. It is mysterious, another thing I do not know.

LIST OF ILLUSTRATIONS

315

LIST OF ILLUSTRATIONS

BIBLIOGRAPHY

Able, Kenneth P., ed., *Gatherings of Angels: Migrating Birds and Their Ecology* (New York, 1999)

Acts of Parliament and Convention from the First Parliament of King James I, 26 May 1424, to the first Parliament of Her Majesty Queen Anne, concluded 25 March 1707 before the Union of the 2 Kingdoms of Scotland and England by Sir James Stewart her Majesty's Advocate (Edinburgh, 1707)

Ascherson, Neal, *Stone Voices: The Search for Scotland* (London, 2002)

Avian Brain Nomenclature Exchange: http://www.avianbrain.org

Bear, Mark F., Connors, Barry, and Paradiso, Michael, eds., *Neuroscience: Exploring the Brain* (Philadelphia, 2001)

Bekoff, Mark, *Minding Animals: Awareness, Emotions and Heart* (Oxford, 2002)

Birkhead, Tim, *The Magpies: The Ecology and Behaviour of Black-Billed and Yellow-Billed Magpies* (London, 1991)

Blunt, William, *Linnaeus: The Compleat Naturalist* (London, 2004)

Boehrer, Bruce Thomas, *Parrot Culture: Our 2500-year-long Fascination with the World's Most Talkative Bird* (Philadelphia, 2004)

Brainard, Michael S., and Doupe, Alison J., *What Songbirds Teach Us About Learning, Nature*, Vol. 417, 16 May 2002

Capote, Truman, *A Capote Reader* (London, 1987)

Carmichael, Alexander, *Carmina Gadelica, Hymns and Incantations from the Gaelic* (London, 1992)

Carson, Rachel, *Under the Sea Wind* (New York, 1941)

Carson, Rachel, *Silent Spring* (London, 1962)

Cennini, Cennino d'Andrea, *The Craftsman's Handbook* (New York, 1960)

Cokinos, Christopher, *Hope is the Thing With Feathers: A Personal Chronicle of Vanished Birds* (New York, 2000)

Coombes Robert A. H., *Mountain Birds* (London, 1952)

Coombs, Franklin, *The Crows* (London, 1978)

Critchley, E. M. R., ed., *The Neurological Boundaries of Reality* (London, 1994)

Darling, F. Fraser, *The Highlands and Islands* (London, 1989)

Deakin, Roger, *Wildwood: A Journey Through Trees* (London, 2007)

Dillard, Annie, *Pilgrim at Tinker Creek* (London 1975)

Edelman, Gerald M., *Bright Air, Brilliant Fire: On the Matter of the Mind* (London, 1992)

Edelman, Gerald M., *Consciousness: How Matter Becomes Imagination* (London, 2000)

Emery, Nathan J., and Clayton, Nicola, *Evolution of the Avian Brain and Intelligence, Current Biology*, Vol. 15, Issue 23, 6 December 2005

Emery, Nathan J., *Cognitive Ornithology: The Evolution of Avian Intelligence, Philosophical Transactions of the Royal Society (Biology)*, 29 January 2006

Feher-Elston, Catherine, *Ravensong* (Flagstaff, AZ, 1991)

Fermor, Patrick Leigh, *A Time of Gifts* (London, 1977)

Halle, Louis J., *The Water Rail* (Leiden, 1963)

Halle, Louis J., *Spring in Washington* (London, 1988)

Halle, Louis J., *Appreciation of Birds* (London, 1989)

Hansell, P. and J., *Doves and Dovecotes* (Bath, 1988)

Heinrich, Bernd, *In a Patch of Fireweed* (London, 1984)

Heinrich, Bernd, *Ravens in Winter* (London, 1990)

Heinrich, Bernd, *Mind of the Raven* (New York, 1999)

http://www.starlingtalk.com/mozart.htm

Hunter, James, *On the Other Side of Sorrow: Nature and People in the Scottish Highlands* (Edinburgh, 1995)

Jarvis, Erich D., 'Evolution of Vocal Learning System in Birds and Humans', in *Evolution of Nervous Systems*, Vol. 2 (Oxford, 2006)

Jarvis, Erich D., et al., *Avian Brains and a New Understanding of Vertebrate Brain Evolution, Nature Reviews Neuroscience*, 6 February 2006

Kilham, Lawrence, *The American Crow and the Common Raven* (College Station, TX, 1989)

Kroodsma, D., *The Singing Life of Birds* (New York, 1995)

Krutch, J. W., *The Twelve Seasons* (New York, 1949)

Krutch, Joseph Wood, *The Great Chain of Life* (London, 1957)

Lint, Kenton C., and Lint, Alice M., *Feeding Cage Birds: A Manual of Diets for Aviculture* (London,1988)

Lopez, Barry, *Of Wolves and Men* (New York, 1978)

Lopez, Barry, *Arctic Dreams: Imagination and Desire in a Northern Landscape* (London, 1986)

Lorenz, Konrad, *King Solomon's Ring* (London, 1952)

Lorenz, Konrad, *On Aggression* (London, 1967)

Lovegrove, Roger, *Silent Fields: The Long Decline of a Nation's Wildlife* (London, 2007)

Mackenzie, Donald Alexander, *Wonder Tales from Scottish Myth and Legend* (New York, 1917)

MacLean, Sorley, *From Wood to Ridge: Collected Poems* (Edinburgh, 1990)

Marchand, Leslie A., ed., *Lord Byron: Selected Letters and Journals* (London, 1982)

Marron, Peter, *A Natural History of Aberdeen* (Aberdeen, 1982)

Marzluff, John M., and Angell, Tony, *In the Company of Crows and Ravens* (New Haven, CT, 2005)

Matthiessen, Peter, *The Birds of Heaven: Travels with Cranes* (London, 2002)

McNeill, F. Marian, *The Scots Kitchen* (Edinburgh, 1993)

Morgan, Edwin, *Collected Poems* (Manchester, 1990)

Moss, Stephen, *A Bird in the Bush: A Social History of Birdwatching* (London, 2004)

Nelson, Richard K., *Make Prayers to the Raven: Koyukon View of the Northern Forest* (Chicago, 1983)

Newton, Michael, *A Handbook of the Scottish Gaelic World* (Dublin, 2000)

New Yorker, the, *The New Yorker Book of Poetry* (New York, 1969)

Nicholson, E. M., *Birds and Men* (London, 1951)

Nottebohm, Fernando, *The Neural Basis of Birdsong*, *Public Library of Science*, May 2005

Oliver, Mary, *Blue Pastures* (New York, 1991)

Ostrom, J., *Archaeopteryx and the Origin of Birds*, *Biological Journal of the Linnaean Society*, 1976

Pliny the Elder, *Natural History: A Selection* (London, 1991)

Rothenberg, David, *Why Birds Sing: A Journey into the Mystery of Bird Song* (London, 2005)

Sax, Boria, *Crow (Animal)* (London, 2003)

Savage, Candace, *Bird Brains: The Intelligence of Crows, Ravens, Magpies, and Jays* (Vancouver, 1995)

Savage, Candace, *Crows: Encounters with the Wise Guys* (Vancouver, 2005)

Scott, John, *Songs of Love and Protest* (London, 1972)

Shaw, Philip, and Thompson, D. B. A., eds., *The Nature of the Cairngorms: Diversity in a Changing Environment* (Edinburgh, 2006)

Shepherd, Nan, *The Living Mountain* (Aberdeen, 1984)

BIBLIOGRAPHY

Spaul, Jo, *Urban Birds* (Oldham, 1999)
Stringer, Chris, *Homo Britannicus: The Incredible Story of Human Life in Britain* (London, 2006)
White, T. H., *The Book of Beasts* (London, 1956)

ACKNOWLEDGEMENTS

Many people have, in one way or another, helped me in the writing of this book.

I would like to thank Helen Ruthven, Mike Angove, Dot McMurray, Willie Primrose, Annie Malcolm, Anne Macleod, Sian Preece, Rosie Mackay, Gilian Dawson, Dorothy Schofield, Olivia McMahon, Janet Chisholm, Zanna Fawcett, Dawn McMillan. Chris and Pauline Robinson extended, as ever, warm hospitality for which I'm grateful, as I am for a memorable day in the hills. For every disputation and for much more, Neil Skinner; for their friendship and for the magpie, Elizabeth and William Johnstone; for his exceptional thoughtfulness towards Chicken, and indeed towards me, John Webb; my cousin and fellow goose-hunter, Roger Alexander, for being the best possible company while not finding geese; Ian Murray, for being an unfailing source of advice, opinion, humour and wisdom.

Vicky Dawson and Arthur Kerr have extended their vast knowledge and expertise to me and have been a source of strength and encouragement – I thank them as I do Jane Stewart, Jaclyn Petrie, Bridget

Verspoor, Ross White, Niall Slater, Paul Traynor and Andy Ma, whose company and enthusiasm transformed many Thursdays, and my colleagues Alasdair Fraser and Heather Gibson who have been generous in every way with their time, encouragement and professional judgement. To Esther Tyldesley of Edinburgh University's Department of Chinese, and to the Hunterian Museum, Glasgow, many thanks for their help.

My agent Jenny Brown, I thank for everything, for her resolution, her kindness and unflagging enthusiasm, for being more constellation than mere star. I am indebted to Sara Holloway for her insights, her tireless work and her enthusiasm – I owe her many and profound thanks. I am grateful too to Amber Dowell and everyone at Granta whose expertise has made this book what it is.

I have drawn on the hard work and knowledge of many people who have written so eloquently on the subject of birds and have found their extraordinary scholarship, their observation and their love of birds a source of inspiration – I am grateful to them all as I am to Helen Macdonald for her beautiful illustrations and for her extraordinary understanding of birds.

And my family, Leah – Queen of St Mark's pigeons – David, Rebecca, Hannah, Ian – to all of them, my gratitude and my love.

INDEX